WOODLAND STEWARDSHIP

A PRACTICAL GUIDE FOR MIDWESTERN LANDOWNERS

Melvin J. Baughman
Alvin A. Alm
A. Scott Reed
Thomas G. Eiber
Charles R. Blinn

MINNESOTA EXTENSION SERVICE

UNIVERSITY OF MINNESOTA

Woodland Stewardship is a cooperative effort of the Minnesota Extension Service, the Minnesota Department of Natural Resources, and the Forest Stewardship Program.

Produced by the Educational Development System, Minnesota Extension Service.

Printed on recycled paper with a minimum of 10% postconsumer waste.

To order additional copies of this book, contact: Minnesota Extension Service Distribution Center, University of Minnesota, 20 Coffey Hall, St. Paul, MN 55108. (612) 625-8173.

ISBN 0-9623116-6-9

Printed in the United States of America

ABOUT THE AUTHORS

Melvin J. Baughman

is an extension specialist—forest resources in the Department of Forest Resources at the University of Minnesota. He develops educational materials and conducts applied research aimed at helping private woodland owners better meet their ownership objectives. Areas of emphasis include economics, policy analysis, and hardwood management. He has worked more than eighteen years in extension forestry and has experience in Minnesota, Pennsylvania, and Kansas. He holds a B.S. in forestry and an M.S. in forest recreation from Michigan State University and a Ph.D. in forest economics and policy from the University of Minnesota.

Alvin A. Alm

is a professor in the Department of Forest Resources at the University of Minnesota. Prior to his retirement in 1992, he taught courses in silviculture at the undergraduate and graduate levels. He has more than twenty-five years of experience as a research project leader with emphasis on site preparation and forest regeneration. He received his Ph.D. degree from the University of Minnesota in 1971, and has worked with the federal government and in Michigan and Maryland as a consulting forester.

A. Scott Reed

is program leader for extension forestry and assistant dean in the College of Forestry at Oregon State University. He formerly was coordinator at the Cloquet Forestry Center and extension specialist—forest resources at the University of Minnesota as well as area forester and research forester with Potlatch Corporation. He has conducted numerous educational and research programs in timber harvesting and economic development. He earned B.S. and M.S. degrees from Michigan State University and holds a Ph.D. in forest economics and policy from the University of Minnesota.

Thomas G. Eiber

is a forest health specialist with the Minnesota Department of Natural Resources. Before joining the DNR in 1987, he taught forest protection at Lakehead University in Thunder Bay, Ontario, for thirteen years. His experience also includes five years of insect evaluations with the U.S. Forest Service and a year as a city forester. He holds B.S. and M.S. degrees in forestry from Michigan State University and a Ph.D. in forest resources from the University of Maine. His current work involves the application of nonchemical technology to control forest pests and the use of geographic information system technology in risk-rating forests.

Charles R. Blinn

is an extension specialist—forest resources in the Department of Forest Resources at the University of Minnesota. He develops educational materials and programs and conducts applied research oriented towards natural resource professionals. His areas of emphasis include forest management, timber harvesting, microcomputer applications, and financial analysis. His professional experience includes industrial research as well as nine years at the University of Minnesota. He earned his Ph.D. in industrial forestry operations from Virginia Tech.

CONTENTS

FEDERAL INCOME TAXES 119

12

FINANCIAL ANALYSIS OF WOODLAND INVESTMENTS 135

13

APPENDIX A: SOURCES OF FORESTRY ASSISTANCE 153

APPENDICES A & B

APPENDIX B: FORESTRY MEASUREMENTS AND CONVERSIONS 159

REFERENCES,
GLOSSARY, INDEX

ACKNOWLEDGMENTS

Publishing funds were provided by the University of Minnesota's Minnesota Extension Service (MES) and the Minnesota Department of Natural Resources, Division of Forestry (DNR). MES funds were allocated from its Renewable Resources Extension Act account, which originates from the United States Department of Agriculture, Extension Service. DNR funds were allocated from the Forest Stewardship Program account, which originates from the United States Department of Agriculture, Forest Service, Northeastern Area State and Private Forestry.

A steering committee recommended content for the book. Members included: Melvin J. Baughman, Department of Forest Resources, and James R. Kitts, Department of Fisheries and Wildlife, College of Natural Resources, University of Minnesota; Christopher R. Brokl, Minnesota Forestry Association; Bernard Dickerson, retired from State and Private Forestry, U. S. Forest Service; Gerald L. Jensen, Robert Tomlinson, and Allen Wickman, Division of Forestry, Minnesota Department of Natural Resources; and A. Scott Reed, Oregon State University.

Many individuals provided content suggestions. We thank Thomas Hovey, a former graduate student in the Department of Forest Resources at the University of Minnesota, for searching the forestry literature for reference materials. Several individuals reviewed early drafts of this publication. They include LeRoy C. Johnson, field representative, Northeastern Area State and Private Forestry, U.S. Forest Service; Ron Overton, regeneration specialist, Northeastern Area State and Private Forestry, U.S. Forest Service; and Sue Brokl, private forest management forester, Division of Forestry, Minnesota Department of Natural Resources.

Besides the authors there were several people with technical skills who were directly involved in producing the book. Mary Ann Hellman in the University of Minnesota's Department of Forest Resources spent countless hours in word processing and coaching some of the authors in computer use. Our freelance editor, Mary Hoff, provided outstanding guidance in blending the authors' writing styles and simplifying technical information. Original artwork, page layout, and overall book design was developed by DRAWN BY DESIGN. All cover design is the work of Deb Thayer, graphic designer for the Minnesota Extension Service. Gail Tischler from the Minnesota Extension Service provided initial consultation on expenses, production steps, and marketing. Richard Sherman from the Minnesota Extension Service served as product manager to estimate expenses, write contracts, and coordinate the editing, design, and printing process. We could not have produced this book without the expert assistance from these people.

The authors greatly appreciate the constructive input by these many individuals, but the authors accept responsibility for the final product, including any oversights or errors.

PREFACE

This book was produced in response to the Minnesota Forest Resource Management Act of 1982, Sec. 18, Subdivision 3, which states that, "the Commissioner of Natural Resources shall prepare and distribute a forest management manual, stressing the concept of multiple use and education and management concerns for small landowners who own at least ten acres of woodlands. The manual shall be prepared with the assistance and cooperation of the University of Minnesota's [Minnesota] Extension Service, Agricultural Experiment Station and College of [Natural Resources] and other public and private forestry organizations."

While intended for use in Minnesota, this material is generally applicable throughout the Midwest. It focuses on growing and regenerating trees for wood products and wildlife while protecting water quality and aesthetic values.

This book will give you, the woodland owner, the background you need to develop and carry out a woodland stewardship plan. It is not a substitute for the years of education and experience held by foresters and other natural resource professionals. For this reason, we encourage you to enlist the help of a forester and seek out other educational materials and programs. Sources of technical assistance are described in Appendix A. There is a glossary of forestry terms beginning on page 177. Additional publications are listed at the end of some chapters and near the end of the book on page 175.

1

PREPARING A WOODLAND STEWARDSHIP PLAN

What will you do with your woodland? As someone who owns forested land, you have a decision to make. You can do nothing. You can occasionally do things that generate income or improve the property's appearance. Or, you can become a woodland steward by actively managing your land for personal benefits, while protecting the quality of its natural resources (soil, water, wildlife, trees, and other plants) for future generations.

Woodlands are a renewable resource, but they require many years to mature. Decisions you make now about timber harvesting, tree planting, or pest control can influence the character of your woodland for the next century. In managing a woodland, you need to plan for the long term because whatever you do—or don't do—will have long-term impacts.

A woodland stewardship plan will help you determine your personal objectives, manage efficiently, avoid costly errors, make knowledgeable decisions, and evaluate your progress.

This chapter describes how you can create a plan for your woodland. You will need to work with a forester to carry out certain steps.

STEP 1: DECIDE WHAT YOU WANT

The first step in planning how to manage your woodland is to develop a list of objectives. What do you want from your woodland? How much do you want? When do you want it? Your management choices will be clearer if your objectives are specific. For example, "to improve the land for wildlife" may be too vague of an objective to guide you toward sound decisions. On the other hand, an objective "to increase the number of grouse on the property" may lead to some very specific management practices.

When you have multiple objectives, be sure to set priorities. Some objectives will be compatible given your resource base, but others may be incompatible, and often only one objective can be maximized. You may not be able to develop realistic objectives until you learn more about the capability of your woodland by conducting an inventory (Step 2).

STEP 2: FIND OUT WHAT YOU HAVE

The second step is to have a forester inventory your woodland to determine what resources you have. Since a forest is dominated by trees, an inventory usually assesses the tree species composition, stand density and age, and tree diameters, heights, quality, and growth rates. Other resources also can be inventoried depending on your objectives. Working with a forester or other natural resource specialist, you can expand your inventory to assess wildlife and fish habitat or other renewable natural resources. For example, the inventory can identify important sites for wildlife breeding, nesting, water, food, and cover.

Although your woodland is just one part of a broader landscape, cumulative effects of management decisions by you and other landowners can greatly alter the landscape over time. Thus as part of the inventory process, you should identify land uses on property that adjoins yours and find out what plans your neighbors have for managing their land. This will better enable you to evaluate the impact that your woodland management activities may have on the landscape. Coordination among neighbors can produce a landscape that meets individual landowner objectives without adversely affecting the environment.

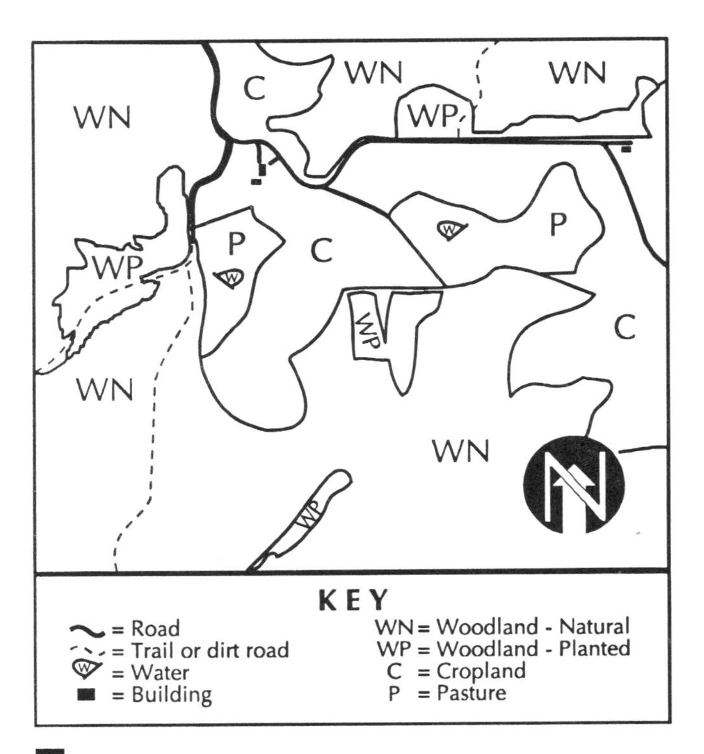

Fig. 1. A base map shows land uses.

Your inventory results will come in handy in several ways. A forester can use them, along with your objectives list, to advise you about alternative management practices and their consequences. An inventory also may help you report and minimize your federal income taxes if it is conducted close to the time when you first acquire the woodland.

Before you begin an inventory, accurately locate and clearly mark property boundaries. Boundaries can be marked with a fence, paint marks on trees, rock piles, stakes, or other means. Clear brush from your property lines to avoid trespass when you or your neighbors carry out forestry practices. If boundaries are not clearly identifiable, you may want to have your land surveyed.

Next, draw one or more maps of the property (Figure 1), approximately to scale, showing the following:

- Property boundaries.
- Woodland boundaries.
- Land uses.
- Roads, trails.
- Utility wires, pipelines, or other rights-of-way.
- Buildings.
- Water resources.
- Unique natural, historical, or archeological resources.

This map will help you and your forester locate woodland accesses and important resources that influence woodland management.

Aerial photographs are especially helpful as a foundation for the map (Figure 2). They usually are available from local offices of the U.S. Department of Agriculture (USDA) Agricultural Stabilization and Conservation Service.

If the property is large and hilly, topographic maps may help you assess slope and aspect as they relate to woodland access and tree growth (Figure 3). Topographic maps are produced and sold by the U.S. Geological Survey, Map Distribution, Federal Center, Building 41, Box 25286, Denver, CO 80225. They also may be available in book, map, or outdoor stores.

Fig. 2. An aerial photograph helps identify land uses.

Fig. 4. A soil type map.

Soil information can help you determine the suitability of your land for different tree species, road or building sites, or other land uses. Soil type maps (Figure 4) and interpretive information about them (Figure 5) may be available from local USDA Soil Conservation Service offices or your local forester.

Gather facts concerning previous land use or management activities that could have influenced the development of your woodland. Such activities might include livestock grazing, agricultural cropping, timber harvesting, tree planting, fires, and pest outbreaks. Foresters use information about these events and their timing to analyze the development of existing woodlands and to predict the results of future management practices.

During the inventory, the forester will prepare a map that separates the woodland into individual stands (Figure 6). Each stand will be an area of approximately 2 to 40 acres that is relatively uniform in tree species composition, tree size distribution, number of trees per acre (stocking), and site quality. Each stand is a management unit and cultural practices are carried out more or less uniformly within a stand.

A more detailed description of a timber inventory is presented in Chapter 2.

Fig. 3. A topographic map has contour lines that show elevation changes.

STEP 3: IDENTIFY POTENTIAL MANAGEMENT PRACTICES

After you identify your objectives and have a forester inventory your woodland, consider all reasonable management practices that would help you meet your objectives.

They might include:

- Planting trees.
- Improving the timber stand (thinning, weeding, culling, pruning).
- Harvesting timber.
- Fencing.
- Improving wildlife habitat.
- Installing erosion control structures on roads.
- Constructing access roads.
- Developing trails.
- Developing recreational facilities.
- Establishing fire protection or controlled burning measures.
- Controlling pests (insects, diseases, animals).
- Controlling weeds and brush.

Seek professional advice on which practices are appropriate for your woodland.

STEP 4: ASSESS LABOR AND FINANCIAL RESOURCES

Once you have developed a list of potential management practices that would help you reach your objectives, evaluate your labor and financial resources available to carry them out. Assess your ability and interest in various forestry practices. Consider how much time you are willing to devote to woodland management, when that time is available, and how long you plan to own the woodland. What is the availability, cost, and quality of contract labor? Consider your financial situation—available capital, cash flow requirements, planning period, rate of return you would like to earn on invested funds, and need for income or products from the woodland.

Finally, assess the availability of needed equipment, facilities, and materials. All of these factors will influence what you can do in your woodland. Chapter 13 provides more detailed information about how you can evaluate the financial efficiency of alternative investments.

MAP SYMBOL & SOIL NAME	EROSION HAZARD	EQUIPMENT LIMITATION	SEEDLING MORTALITY	WINDTHROW HAZARD	PLANT COMPETITION	COMMON TREES	SITE INDEX	TREES TO PLANT
457G LaCrescent 45-70% slope	severe	severe	slight	moderate	moderate	northern red oak white oak American basswood	55 55 55	eastern white pine, white oak, American basswood, northern red oak, white ash
580B, 580C2 Blackhammer-Southridge 3-12% slope	slight	slight	slight	slight	moderate	northern red oak American basswood white oak shagbark hickory	70 70 62 60	northern red oak, American basswood, sugar maple, white oak, eastern white pine, white ash, red pine
580D2 Blackhammer-Southridge 12-20% slope	moderate	moderate	slight	slight	moderate	northern red oak American basswood white oak shagbark hickory	70 70 62 60	northern red oak, American basswood, sugar maple, white oak, eastern white pine, red pine, white ash
584F Lamoille 30-45% slope	severe	severe	moderate	moderate	moderate	northern red oak American basswood green ash white oak shagbark hickory sugar maple	58 58 52 52 50 50	northern red oak, white oak, American basswood, eastern white pine, white ash
586C2 Nodine-Rollingstone 4-12% slope	slight	slight	slight	moderate	moderate	northern red oak white oak shagbark hickory American basswood sugar maple	65 60 60 70 60	northern red oak, white oak, American basswood, sugar maple, eastern white pine, white ash
586D2 Nodine-Rollingstone 12-20% slope	moderate	moderate	moderate	moderate	moderate	northern red oak white oak American basswood sugar maple	65 60 70 60	northern red oak, American basswood, sugar maple, eastern white pine, white oak, white ash
592E Lamoille 20-30% slope	moderate	moderate	moderate	moderate	moderate	northern red oak American basswood green ash sugar maple	55 55 52 50	northern red oak, white oak, American basswood, eastern white pine
592E Elbaville 20-30% slope	moderate	moderate	slight	slight	moderate	northern red oak white oak American basswood sugar maple black walnut	65 60 65 65 65	northern red oak, black walnut, eastern white pine, white oak, sugar maple, American basswood

Fig. 5. Typical soil interpretation for woodland management. (From U. S. Soil Conservation Service. 1984. Soil Survey

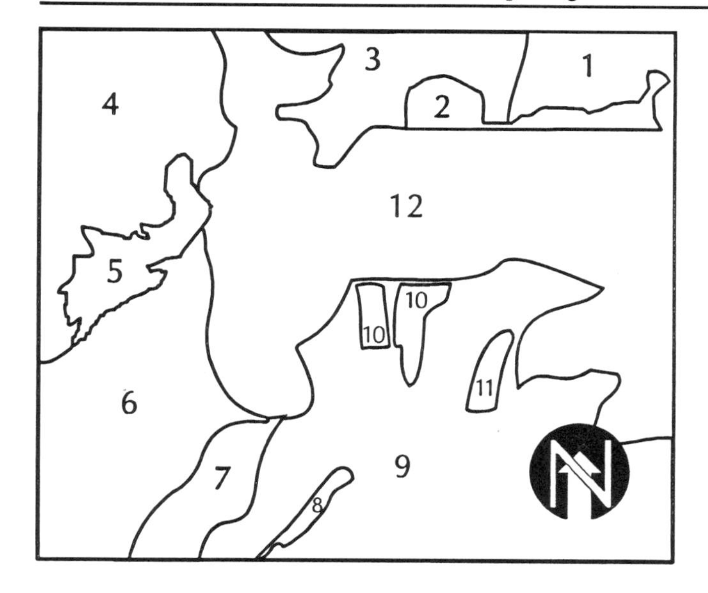

STAND NO.	DESCRIPTION
1	Red and white oak, basswood, sugar maple; 6- to 12-inch diameter; needs timber stand improvement.
2	Old field planted with red oak; 1 yr. old; let grow.
3,9	Red and white oak; mixed sizes; needs group selection harvest and timber stand improvement.
4,6	Red and white oak with pockets of aspen on upland; 10- to 16-inch diameter; let grow another 10 to 15 yrs., then harvest.
5	Red pine plantation; 26 yrs. old; 6- to 12-inch diameter; thinned recently; let grow.
7,11	Red and white oak, basswood, and aspen; 5-yr.-old natural regeneration resulting from clearcut. Let grow 2 to 5 yrs. then release oaks from competition and thin stump sprouts.
8	White pine plantation; 15 yrs. old; 4- to 8-inch diameter; let grow 5 to 10 yrs., then thin.
10	Red and white pine plantation; 20 yrs. old; 1- to 8-inch diameter; needs thinning.
12	Cropland and pasture.

Fig. 6. A timber stand map.

STEP 5: DEVELOP AN ACTIVITY SCHEDULE

Next, prepare an activity schedule that lists management practices and approximately when you expect to perform them. This schedule should cover at least five to ten years. If your woodland is large—perhaps several hundred acres—activities may occur every year. If it is smaller, management activities will occur less often, perhaps only once every ten years. Regardless of its size, inspect your woodland at least annually. Walk through the woodland and look for damage by pests, fire, or wind; unauthorized harvest; damaged fences; and soil erosion.

STEP 6: KEEP GOOD RECORDS

It will be difficult to update your plans and make sound decisions about the future unless you keep accurate records of what you have done. Records also will be important when filing income tax reports and perhaps for settling an estate.

Management records may include:

- Management plan.
- Timber inventory.
- Management activities accomplished (what, when, where).
- Sources of forestry assistance (name, address, telephone).
- Association memberships.
- Suppliers of materials and equipment.
- Contracts.
- Insurance policies.
- Forestry income and expenses.
- Deeds, easements.

2

CONDUCTING A WOODLAND INVENTORY

This section describes basic inventory procedures to help you understand and interpret inventory information. Your woodland should be inventoried by a forester because of the potential for measurement errors by untrained individuals. With training, practice, and the right tools, however, you could learn to assist a forester with the inventory.

A woodland inventory usually involves measuring timber volume and stand density. It also may involve measuring tree growth rate and tree quality. This information can help a forester assess the need for thinning, harvest, or regeneration. It also could be used to estimate the timber's market value and project financial returns over time.

Because woodlands often are viewed as a source of timber, an inventory usually focuses on assessing trees as potential wood products. Information obtained during the inventory, however, provides a snapshot of your woodland that is equally valuable for assessing wildlife habitat, planning access roads or trails for recreation, and understanding the quality of your soil and water resources.

An inventory should focus on the resources most important to you. Since an inventory involves some time and expense, be sure to determine what information you really need before beginning it.

INDIVIDUAL TREE VOLUMES

For wood products the most useful and highest valued part of a tree is the main stem, also called the trunk or bole. To estimate the wood volume of a tree's stem, a forester measures the diameter and merchantable height and then finds the corresponding volume in a table.

Tree Diameter

Tree diameter is measured on the main stem 4-1/2 feet above ground. This is referred to as diameter at breast height (DBH). DBH usually is measured to the nearest inch. A steel diameter tape, calibrated to permit direct tree diameter readings, frequently is used to measure DBH.

To measure tree diameter with a diameter tape, wrap the tape around the tree at breast height, perpendicular to the lean of the tree, standing on the uphill side of the tree (Figure 7a). If there are branches or other protrusions at 4-1/2 feet, place the tape at the first unobstructed location above them (Figure 7b). If the tree forks below 4-1/2 feet, consider each stem to be a separate tree and record separate measurements for each (Figure 7c).

Diameter tapes can be purchased from forestry supply companies or you can make your own tape from any narrow, flexible banding material such as steel, plastic, or cloth. Write or inscribe marks indicating inches of tree diameter every 3.14 inches along the tape. If a diameter tape is not available, measure tree circumference by wrapping a normal tape measure around the tree and dividing the resulting circumference by 3.14 to determine DBH.

Tree Height

Total tree height is the height from the ground to the top of a tree. Not all of the wood in a tree, however, is merchantable. Merchantable tree height is the height of the stem from the top of the expected stump to the upper limit of utilization in the tree. Stump height normally is 6 inches for softwood species and 12 inches for hardwood species, but varies depending on what table you use to determine volume. The upper limit of utilization is where the main stem reaches a minimum usable top diameter, a main fork, or a serious defect such as a hole or a point of decay, or where excess limbs occur.

Standards for merchantability vary widely depending on local product markets. The usual minimum top diameter inside the bark (DIB) for sawlogs and pulpwood is the larger of 50 percent of tree DBH, or 8 inches for sawlogs and 4 inches for pulpwood. Veneer trees must have a top DIB of at least 10 inches. Merchantable tree height usually is measured in 8-foot lengths called half-logs or bolts, but may be measured to the nearest 2 feet on high-value trees.

Tree heights often are estimated with a clinometer or Merritt hypsometer—instruments that can be purchased from forestry equipment dealers. You can make a simple hypsometer by placing marks at 4-inch intervals along a stick or lath. Each mark represents one 8-foot bolt when the stick is used as follows (Figure 8):

1. Stand 50 feet from the tree center in a direction such that the tree does not lean toward or away from you.
2. With the stick in hand, extend your arm out 25 inches from your eye. Hold the stick vertically and in line with the trunk of the tree being measured.
3. Find the upper limit of utilization. Remember that point.
4. Raise or lower the stick as needed until you can sight along the bottom of the stick to stump height. Then, **moving your eyes, not**

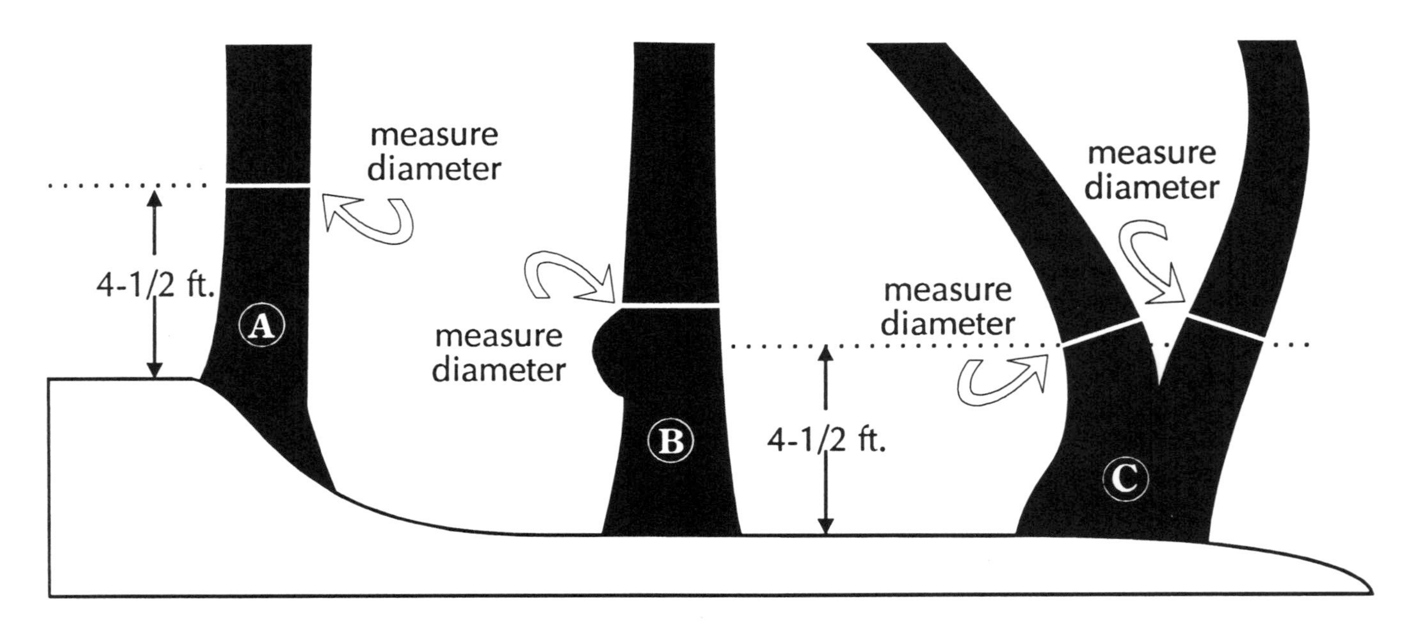

Fig. 7. Measure stem diameter at breast height (DBH).

your head, look up and read the stick measurement that corresponds to the upper limit of utilization point. Count the number of bolts between the stump and the merchantable height and record this number.

If you are unable to get a good sight on the tree from a distance of 50 feet, stand 25 feet from the tree and divide the resulting height measurement by two.

Tree Defects

The main stem is the most useful part of a tree for conventional wood products such as lumber, pulpwood, posts, and poles. Many tree stems contain defects that make portions of them unusable for certain products. The location and severity of defects will influence the volume of usable wood and the types of products that can be cut from a tree.

Defects that may reduce the total volume of usable wood in the tree include decay, forks, sweep, crook, and deep cracks (Figure 9). If the entire tree is unusable because of an excessive amount of defect, it is a cull tree and its volume should not be measured. If a major defect occurs toward the top of the tree, measure the total merchantable height below the defect. If a major defect occurs between stump height and merchantable height, measure total merchantable height and estimate the percent of volume that is defective. Then deduct this from the overall tree volume. Because tree diameter decreases with increasing height in the tree, a defect near the top of the tree will require a smaller deduction than a similar defect near the base.

Defects such as knots, burls, small limbs, insect holes, and bird peck will not affect the total volume of wood, but they will reduce tree quality, or grade (Figure 10). Tree grade influences the types of products that can be made from a tree and therefore the stumpage price. In the Midwest, grading rules are not often formally used to evaluate standing timber except in high-quality hardwood stands.

Estimating defect can be very difficult. If you have many trees with defects or high-value tree species, it is important to have a forester estimate your timber volume and account for the defect.

Tree Volume

Once you have measured DBH and merchantable height, refer to a volume table to estimate the tree's volume. There are different volume tables for different types of wood product.

Sawtimber and Veneer Trees

Trees of sufficient size and quality to produce logs that can be sawed into lumber are referred to

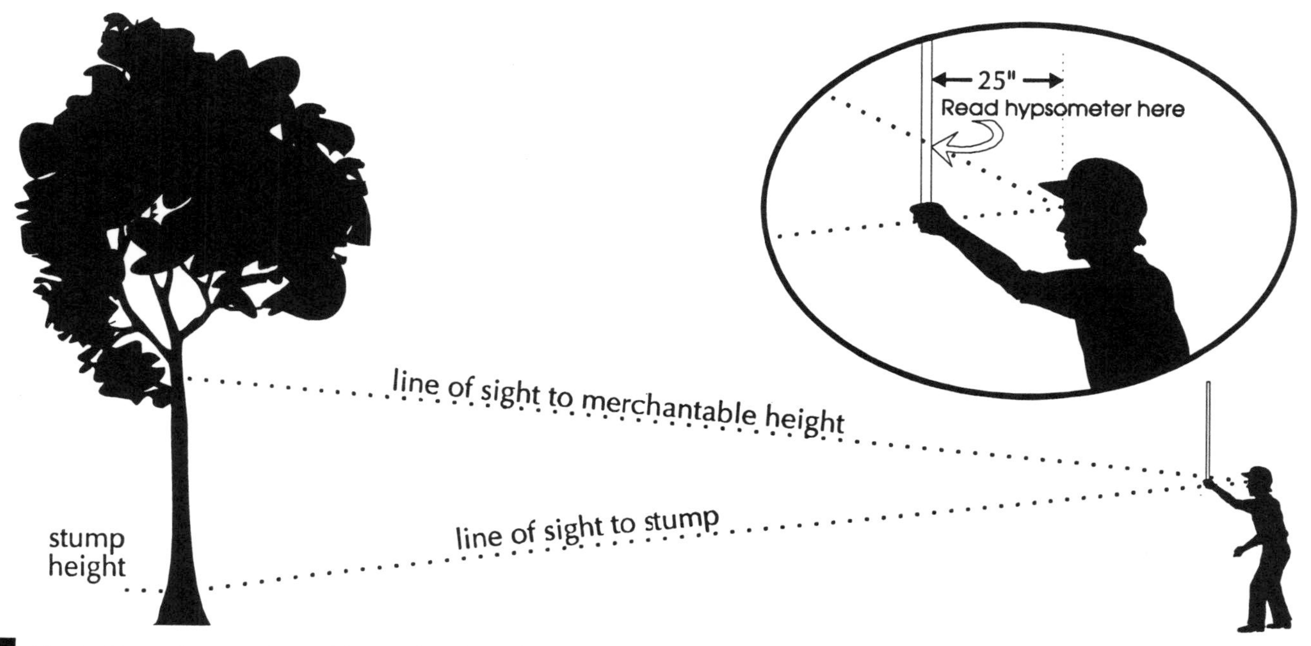

Fig. 8. Measuring merchantable tree height with a hypsometer.

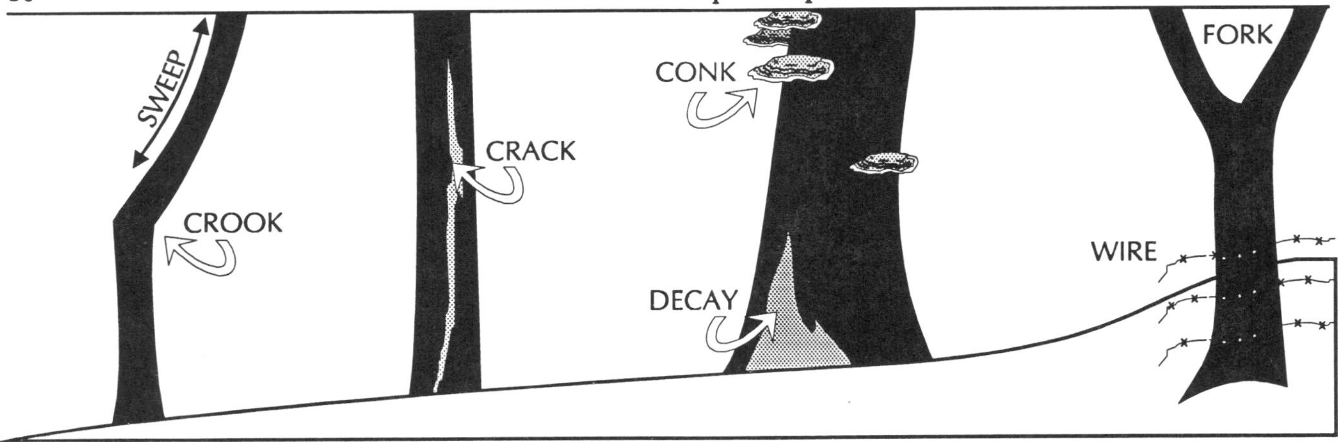

Fig. 9. Tree defects that reduce the total volume of usable wood.

as **sawtimber**. To qualify as sawtimber, trees should have at least one 8-foot bolt, be at least 10 inches DBH, and have a top DIB that is the larger of either 8 inches or 50 percent of DBH. For example, to be a sawlog, a tree of 20 inches DBH should have a minimum top DIB of 10 inches.These specifications are typical, but individual buyers may have different specifications. Sawtimber trees must not contain too many defects such as excessive branchiness, decay, scars, bulges, bark distortions, holes, branch stubs, or crook.

Individual trees of many species (e.g., black walnut, white ash, sugar maple, red oak, white oak) that are of exceptional quality, have at least one 8-foot bolt, are at least 16 inches DBH, and contain bolts that have a top DIB of at least 10 inches often can be sold as **veneer trees**. Logs harvested from these trees will be sliced or peeled into thin sheets. Such trees are more valuable than sawtimber.

The basic unit for estimating volume for both sawtimber and veneer trees is the board foot. A board foot is a piece of wood of any shape that contains 144 cubic inches of wood (e.g., 12 inches x 12 inches x 1 inch, or 6 inches x 6 inches x 4 inches). Timber value is often described in dollars per thousand board feet (MBF).

Many rules have been developed to estimate sawtimber and veneer tree volume. In this book, a **tree rule** will refer to a table that estimates wood volume in a standing tree; a **log rule** will refer to a table that estimates wood volume in a cut log. Tree and log rules estimate the number of board feet of lumber that can be sawed from a tree. These rules differ in their assumptions about factors such as tree taper, board thickness, **kerf** (saw thickness), and minimum and maximum board width. They are never totally accurate because the assumptions upon which they are based seldom occur. In addition, it is difficult to accurately measure volume losses from defects.

The **Scribner rule** (Table 1) and the **Doyle rule** (Table 2) are used most often in the Midwest. On the whole the Scribner rule is more accurate than the Doyle rule. The Doyle rule underestimates the volume of larger trees.

Pulpwood

Several tree species can be sold for **pulpwood.** These trees are converted to chips or fibers used to manufacture products such as paper, hardboard, and structural board. Minimum DBH for pulpwood trees is 5 inches. Minimum DIB is the

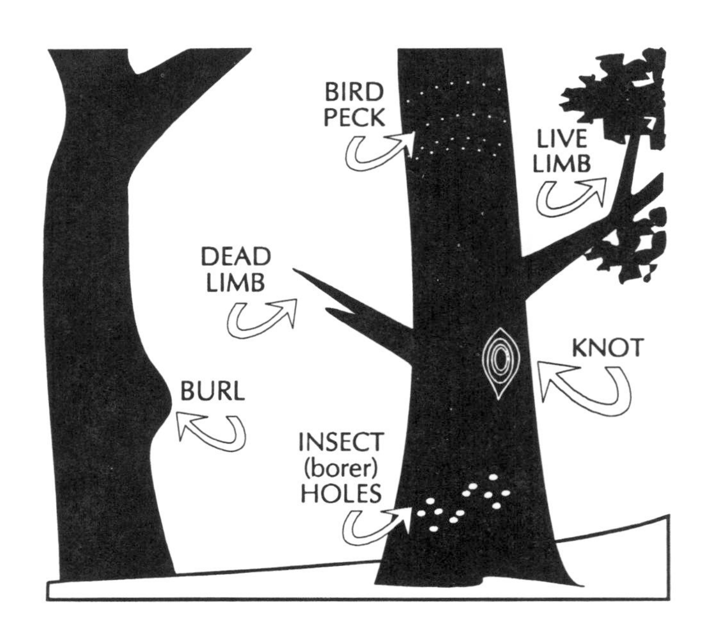

Fig. 10. Tree defects that reduce wood quality.

larger of either 4 inches or 50 percent of tree DBH. (The minimum DIB for a pulpwood tree with a DBH of 12 inches, therefore, is 6 inches.) In the Lake States pulpwood commonly is cut to 100-inch lengths.

The basic unit for estimating pulpwood volume in trees is the **cord**. A standard cord is a closely stacked pile of logs containing 128 cubic feet of wood, bark, and air spaces between logs. A cord frequently is described as a stack of wood 8 feet long, 4 feet high, and 4 feet wide (Figure 11). The solid wood content (excluding bark and air space) of a cord varies from about 65 to 95 cubic feet depending on the diameter, roughness, and crookedness of the pieces. An accepted average value in the Lake States is 79 cubic feet of wood per cord.

Table 3 can be used for estimating pulpwood volume.

! NOTE: Tables 1, 2, and 3 should be used only for standing trees, not cut logs. Cut logs usually are measured by log length and DIB at the small end. Log volume tables differ from those shown here.

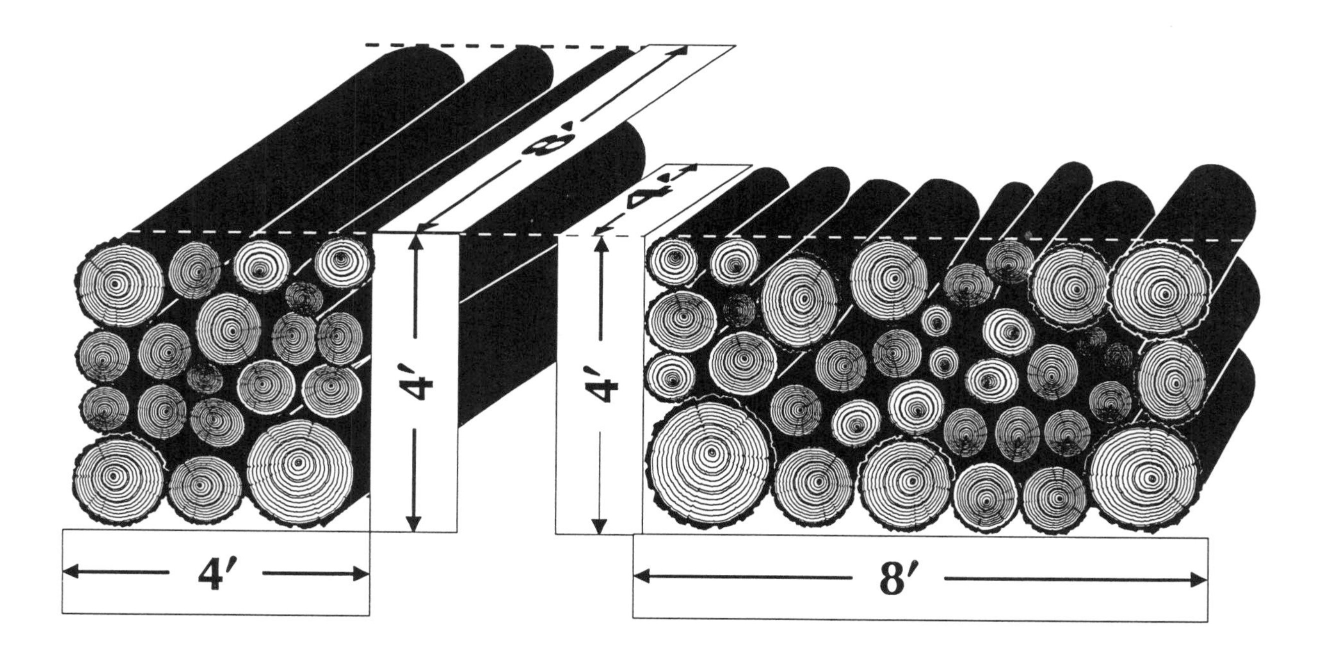

Fig. 11. A standard cord of wood.

Table 1. Tree volume in board feet (Scribner rule).

How To Use Table 1. As an example, a tree with a DBH of 22 inches and 32 feet (four bolts) of merchantable height will yield approximately 286 board feet. If there were 10 percent defect in the tree, total tree volume would be reduced to 257 board feet. A reasonable range of volume estimates per acre is 1,000 to 15,000 board feet.

DBH in inches	NUMBER OF 8-FOOT BOLTS											
	1	2	3	4	5	6	7	8	9	10	11	12
	BOARD FEET[1]											
9	14	—	—	—	—	—	—	—	—	—	—	—
10	16	34	52	65	—	—	—	—	—	—	—	—
11	—	37	58	79	100	117	—	—	—	—	—	—
12	—	42	64	89	114	139	164	181	—	—	—	—
13	—	48	71	97	129	156	185	214	244	—	—	—
14	—	—	80	109	140	177	207	240	273	306	334	—
15	—	—	90	122	156	195	233	268	305	342	380	417
16	—	—	102	136	174	215	260	299	340	382	424	466
17	—	—	118	158	201	249	301	345	393	442	490	538
18	—	—	135	180	230	285	344	395	450	505	560	615
19	—	—	153	205	262	323	390	447	510	572	635	697
20	—	—	177	230	294	364	439	503	573	644	714	784
21	—	—	—	258	329	406	490	562	641	719	797	876
22	—	—	—	286	365	451	545	625	711	799	885	973
23	—	—	—	317	404	499	602	690	786	882	978	1074
24	—	—	—	348	444	548	661	759	864	970	1075	1181
25	—	—	—	381	486	600	724	830	945	1061	1176	1292
26	—	—	—	416	531	655	789	905	1031	1157	1282	1408
27	—	—	—	452	576	711	857	983	1119	1256	1392	1529
28	—	—	—	489	624	770	928	1064	1212	1360	1507	1655
29	—	—	—	528	674	831	1002	1149	1308	1468	1626	1786
30	—	—	—	577	725	894	1078	1236	1407	1579	1750	1922

[1] Board foot volume is assumed to be the gross scale above a one-foot stump to a top DIB that is the larger of either 8 inches or 50 percent of the tree DBH. Volumes outside the tabulated range may be estimated by applying the following formula to an 8-foot bolt and then summing values to provide tree volume estimates:

Bolt volume (board feet) = $0.395d^2 - d - 2$

Where: d = small end inside bark diameter of the bolt, in inches.

SOURCE: Burk, T. E., T. D. Droessler, and A. R. Ek. 1986. *Taper Equations for the Lake States Composite Volume Tables and Their Application.* University of Minnesota, Department of Forest Resources, St. Paul, MN 55108.

Table 2. Tree volume in board feet (Doyle rule).

DBH in inches	NUMBER OF 8-FOOT BOLTS											
	1	2	3	4	5	6	7	8	9	10	11	12
	BOARD FEET[1]											
9	8	—	—	—	—	—	—	—	—	—	—	—
10	9	19	29	37	—	—	—	—	—	—	—	—
11	—	21	33	46	59	69	—	—	—	—	—	—
12	—	25	39	54	70	86	—	—	—	—	—	—
13	—	31	45	62	83	102	121	141	160	—	—	—
14	—	—	53	72	94	120	140	164	187	210	230	—
15	—	—	62	84	108	137	164	190	217	244	272	299
16	—	—	73	97	125	156	190	219	250	282	313	345
17	—	—	88	117	150	187	228	262	300	338	375	413
18	—	—	104	139	178	222	270	310	354	399	443	487
19	—	—	121	162	208	259	315	361	413	465	516	568
20	—	—	145	188	240	299	363	417	476	536	595	655
21	—	—	—	215	275	342	415	476	544	612	680	748
22	—	—	—	244	312	388	471	540	616	693	770	847
23	—	—	—	275	352	436	530	607	693	780	868	952
24	—	—	—	307	393	488	592	679	775	871	967	1063
25	—	—	—	342	437	542	658	755	861	968	1074	1181
26	—	—	—	378	484	600	727	834	952	1070	1187	1305
27	—	—	—	417	533	660	800	918	1047	1176	1305	1435
28	—	—	—	457	584	723	876	1005	1146	1288	1430	1571
29	—	—	—	499	638	789	956	1097	1251	1405	1559	1714
30	—	—	—	551	694	858	1039	1192	1359	1527	1695	1862

[1]Board foot volume is assumed to be the gross scale above a one-foot stump to a top DIB that is the larger of either 8 inches or 50 percent of the tree DBH. Volumes outside the tabulated range may be estimated by applying the following formula to an 8-foot bolt and then summing values to provide tree volume estimates:

Bolt volume (board feet) = $0.395d^2 - d - 2$

Where: d = small end inside bark diameter of the bolt, in inches.

SOURCE: Burk, T. E., T. D. Droessler, and A. R. Ek. 1986. *Taper Equations for the Lake States Composite Volume Tables and Their Application.* University of Minnesota, Department of Forest Resources, St. Paul, MN 55108.

Table 3. Tree volume in rough cords.

How to Use Table 3. As an example, a tree with a DBH of 10 inches and 24 feet (3 bolts) of merchantable height will yield approximately 0.10 cords. If the tree contained 10 percent defect, tree volume would be reduced to 0.09 cords. A reasonable range of volume estimates per acre is 5 to 45 cords.

DBH in inches	NUMBER OF 8-FOOT BOLTS											
	1	2	3	4	5	6	7	8	9	10	11	12
	CORDS[1]											
5	.01	.02	.03	—	—	—	—	—	—	—	—	—
6	.02	.03	.04	.05	.06	—	—	—	—	—	—	—
7	—	.04	.05	.06	.08	.10	.11	—	—	—	—	—
8	—	.05	.06	.08	.10	.12	.14	.15	.17	—	—	—
9	—	.06	.08	.10	.12	.15	.17	.20	.22	—	—	—
10	—	.07	.10	.12	.15	.18	.21	.24	.27	.30	—	—
11	—	.09	.12	.15	.18	.22	.26	.29	.33	.37	.39	—
12	—	—	.14	.18	.22	.26	.31	.35	.39	.44	.48	—
13	—	—	.16	.21	.25	.31	.36	.41	.46	.51	.56	.61
14	—	—	.19	.24	.29	.35	.42	.47	.53	.59	.65	.71
15	—	—	.22	.27	.34	.41	.48	.54	.61	.68	.75	.82
16	—	—	.25	.31	.38	.46	.55	.62	.70	.78	.86	.93
17	—	—	.28	.35	.43	.52	.62	.70	.79	.88	.97	1.05
18	—	—	.31	.40	.49	.59	.69	.78	.88	.98	1.08	1.18
19	—	—	.35	.44	.54	.65	.77	.87	.98	1.10	1.21	1.32
20	—	—	.39	.49	.60	.72	.86	.97	1.09	1.21	1.34	1.46
21	—	—	—	.54	.66	.80	.94	1.07	1.20	1.34	1.47	1.61
22	—	—	—	.59	.73	.88	1.03	1.17	1.32	1.47	1.62	1.77
23	—	—	—	.65	.80	.96	1.13	1.28	1.44	1.61	1.77	1.93
24	—	—	—	.70	.87	1.04	1.23	1.39	1.57	1.75	1.92	2.10
25	—	—	—	.76	.94	1.13	1.34	1.51	1.70	1.90	2.09	2.28
26	—	—	—	.83	1.02	1.22	1.45	1.64	1.84	2.05	2.26	2.47
27	—	—	—	.89	1.10	1.32	1.56	1.76	1.99	2.21	2.44	2.66
28	—	—	—	.96	1.18	1.42	1.68	1.90	2.14	2.38	2.62	2.86
29	—	—	—	1.03	1.26	1.52	1.80	2.04	2.29	2.55	2.81	3.07
30	—	—	—	1.11	1.35	1.63	1.92	2.18	2.45	2.73	3.01	3.28

[1]Volume is standard unpeeled cords and includes the stem wood above a 1-foot stump to a top DIB that is the larger of either 4 inches or 50 percent of the tree DBH. Careful piling of harvested bolts is assumed, equivalent to 79 cubic feet of wood or 92 cubic feet of wood and bark per cord. Volumes outside the tabulated range may be estimated by applying the following formula to each 8-foot bolt and then summing values to provide tree volume estimates:

Bolt volume = $0.0003(d^2 + D^2)$

Where: d = small end inside bark diameter of the bolt, in inches; D = large end inside bark diameter of the bolt, in inches.

SOURCE: Burk, T. E., T. D. Droessler, and A. R. Ek. 1986. *Taper Equations for the Lake States Composite Volume Tables and Their Application.* University of Minnesota, Department of Forest Resources, St. Paul, MN 55108.

TIMBER STAND VOLUMES

To determine the volume of wood in a tract of 5 acres or less, or in a tract that contains highly valuable material such as black walnut veneer, measure each tree individually and total those volumes.

The simplest method for estimating timber volume in a larger area is to estimate the volume of sample plots and then apply an expansion factor based on the area of the stand and the area sampled. A reasonably accurate stand volume may be obtained by measuring the volume of trees in a number of 1/20th-acre circular sample plots (radius 26.33 feet). A suggested rule is to sample one plot per acre in tracts of 30 acres or less. In larger tracts, sample 24 plots plus one additional plot for each 5 acres. For example, in a 37-acre tract you would sample 32 plots (24 + [37/5] = 32 plots). This sampling intensity usually will provide estimates within 20 percent of the actual stand volume.

Sample plots should be distributed throughout the tract—don't just locate them in what appears to be an "average" part of the tract. This is especially critical in tracts in which tree size, species composition, and tree density vary. One way to obtain a broad distribution is to locate plots at equal distances along parallel lines. Using a compass, run lines perpendicular to major landforms, drainages, or changes in timber type (Figure 12).

The following formula can be used to determine the distance in feet between lines and plots:

$$D = 208.71\sqrt{A/n}$$

Where: D = Distance (feet) between lines and sample plots on a line

A = Number of acres in the tract

n = Number of sample plots

As an example, if there are 37 acres in the tract and thirty-two plots are sampled, the distance between lines and sample points along a line would be 224 feet

$$(208.71\sqrt{37/32} = 224)$$

If you accompany a forester during the inventory, you most likely will follow this procedure:

1. Based on topography and timber types, establish a pattern for laying out sample plots.

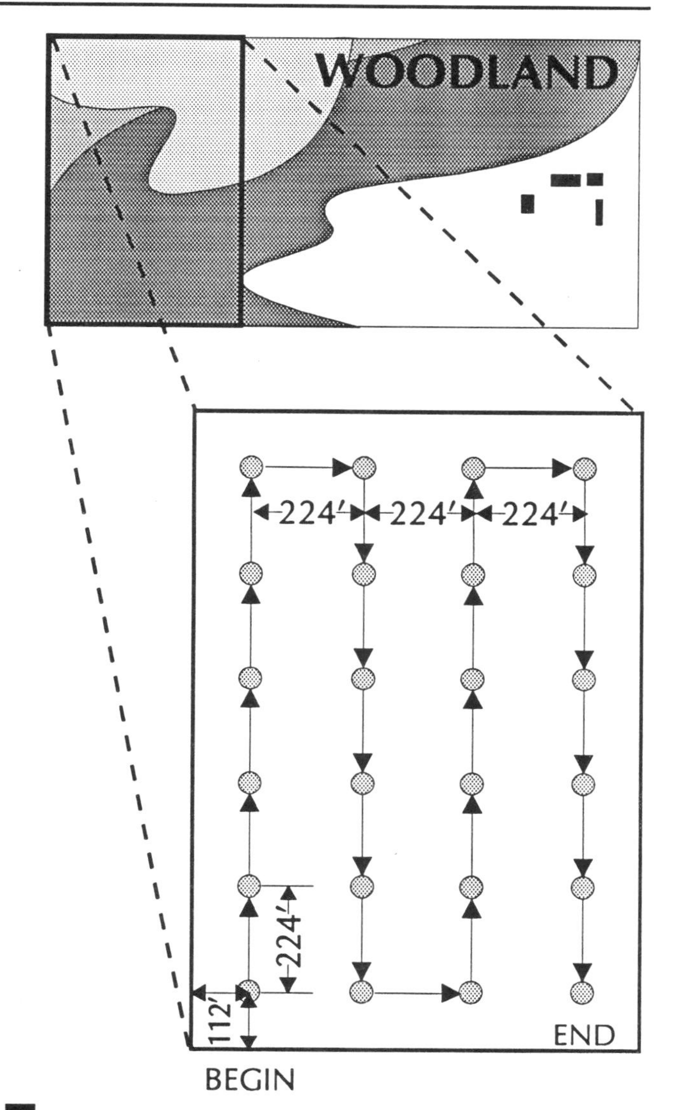

Fig. 12. Distribution of sample plots in a woodland.

2. When sampling a large area, start from a known point, such as a property corner. Use a compass to orient lines through the woodland on which sample plots will be located. Pace or measure with a tape to the first sample plot center and mark it.
3. Identify a nearby natural landmark such as a dead or unusual tree and proceed from there in a clockwise direction, recording measurements for all merchantable trees within 26.33 feet of the plot center. (If possible, one person measures trees while another records information.)
4. Follow the planned compass line for the required distance to the second plot location and measure those trees.
5. Repeat this process for each sample plot.

Fig. 13. A tree grows a new ring of wood each year.

Foresters frequently use aerial photographs in laying out sample plots. Through the use of aerial photographs, the woodlot can be broken down into homogeneous (uniform) forest types. Then fewer sample plots are needed to arrive at a reasonably accurate estimate of volume.

SITE INDEX

As part of a woodland inventory, site quality usually is evaluated to help predict how well trees will grow compared with other sites. The rate that a tree grows depends partly on genetic characteristics and partly on site factors, including soil, moisture, climate, slope, and aspect (direction a slope faces). Rather than measuring numerous site factors, foresters often judge site quality based on the total height that dominant and codominant trees will grow in a given period—usually 50 years in the Midwest. Trees are expected to grow taller on good sites than on poor ones in the same time period. This measure of site quality is called **site index.** Site index curves have been constructed for many tree species so that site quality may be determined for a stand of trees larger than saplings if average tree age and average total tree height are known.

Tree Age

You can determine a tree's age by felling it and counting the annual growth rings on the stump or log (Figure 13). When tree age is to be used for determining site index, the tree to be felled must be a species that is to be favored by management, is common in the stand, and is in a dominant or codominant position in the stand (see Figure 15).

A less destructive method for determining tree age is to take core samples using an increment borer. An increment borer is similar to a hand drill but has a hollow drill bit. It extracts a core of wood about the size and shape of a pencil. This core shows annual rings as bands of light and dark wood (Figure 16). If you take the increment core several feet above the ground, you will need to estimate how many years it took the tree to reach that height and add those years to the count of rings on the core to determine tree age. Increment borers can be bought from forestry equipment suppliers.

A third method for determining age can be used for conifers such as red (Norway) pine, white pine, and balsam fir that produce one whorl of new branches each year near the tree top. You can estimate age of such trees, especially if they are young, simply by counting whorls (Figure 14).

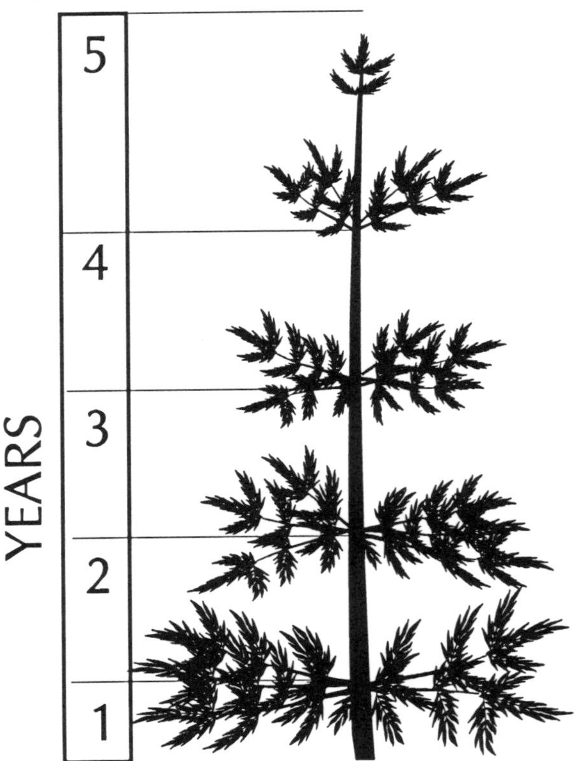

Fig. 14. Count the whorls of branches to determine the age of a conifer.

Determining Site Index

To determine site index, find the average age and average total height of dominant and codominant trees in pure, even-aged stands on the site index curves (an example for red pine is shown in Figure 17). The curve closest to the point you have identified is the site index for that site; it is the height those trees would be expected to reach at a standard index age (in the Midwest, 50 years). A **pure** stand is one in which at least 80 percent of the trees are of the same species. An **even-aged** stand is one where the age difference between the youngest and oldest tree in a stand does not exceed 20 percent of the projected rotation length. A **rotation** is the number of years required to establish and grow trees to a specified size, product, or condition of maturity at which they can be harvested.

For example, suppose you want to find the site index for an even-aged stand of red pine that averages 80 years old and 90 feet tall. Find 80 years on the bottom axis in Figure 17, then follow the vertical line upward from that point until it intersects the 90-foot level on the left vertical axis. Now follow the site index curve from the junction of those two lines to the right vertical axis, where you will find that the site index is 65. These trees would be expected to be 65 feet tall at age 50. The higher the site index, the better the site. Site index curves for common tree species are shown in Appendix C.

Site index curves are reasonably accurate measures of site quality for stands older than 20 years. For younger stands, special equations are used to estimate site index.

BASAL AREA

Basal area (BA) per acre, an index of stand density, is another measurement often made in a woodland inventory. It is defined as the sum of cross-sectional areas of the main stems measured 4-1/2 feet above ground of all trees in an acre (Figure 18). Different forest types and tree ages have different optimal basal areas per acre for best tree growth. If the stand has a high BA for its type and age, it needs a thinning or harvest. General BA guidelines for most forest types can be found in Chapter 6. A forester can tell you what BA is best for your stands.

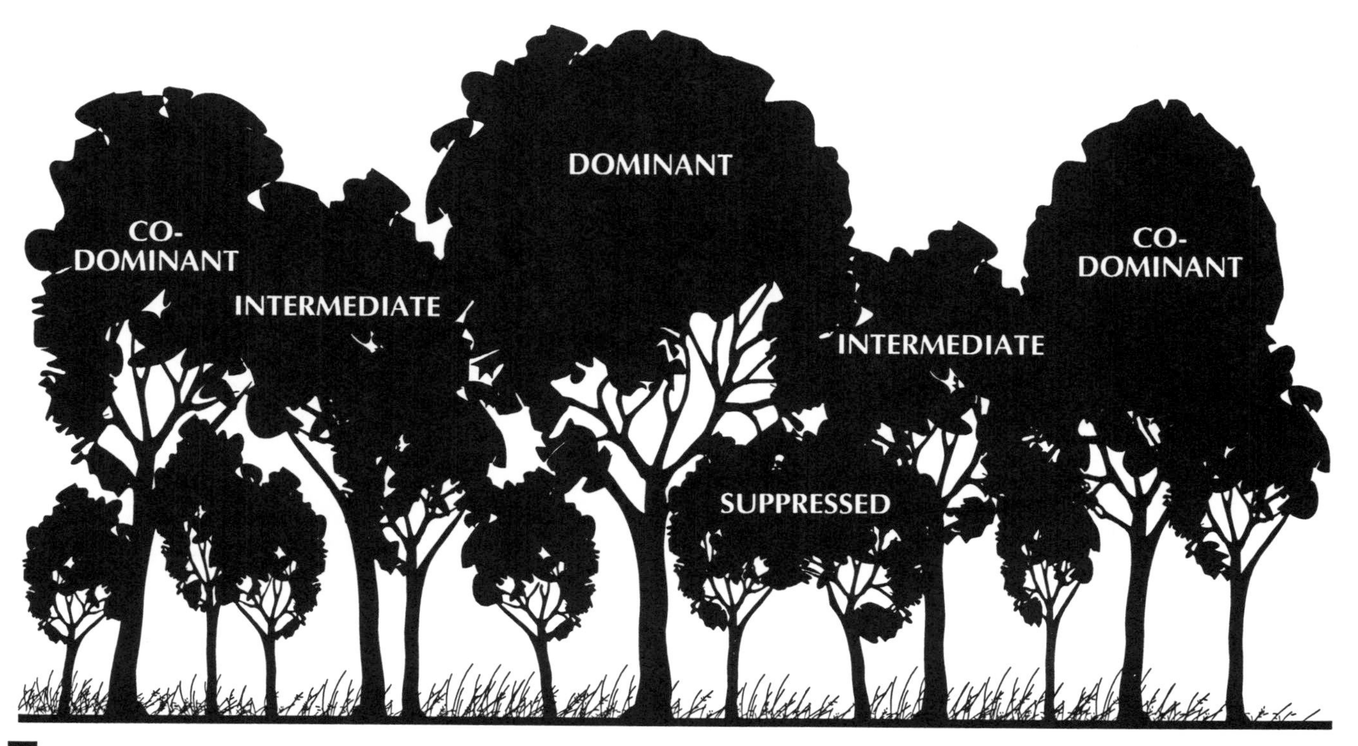

Fig. 15. The dominance of a tree refers to the position of its crown relative to other trees in the canopy. Dominant trees have relatively large crowns and are taller than most other trees in the stand. Codominant trees make up the general canopy level. Intermediate trees are slightly lower than the general canopy level and have relatively small crowns. Suppressed trees are below the general canopy level.

Basal area of a tree can be calculated with the formula:

$$BA = 0.005454\ D^2$$

Where: D = DBH

However, since BA per acre is such a common and useful measure of stand density, tools have been developed to determine it quickly without actually measuring and calculating the basal area of every tree.

A stick-type angle gauge is one such tool. A 10-factor gauge consists of a rod 33 inches long with a piece of metal exactly 1 inch wide attached to one end and protruding above the stick about 1 inch (Figure 19). You could easily make your own, but measurements must be exact. To use the angle gauge, stand in a fixed spot with the "open end" of the stick near your eye and the 1-inch metal intercept pointed at a nearby tree. As you pivot in a circle, count all the trees with a DBH that appears larger than the 1-inch metal intercept. Your eye always should be over the fixed point. Each tree you count represents 10 square feet of basal area per acre, hence the "10-factor" gauge. If you count twelve trees from a fixed point that all have diameters that appear larger than the metal intercept, your stand density at that point is 120 square feet

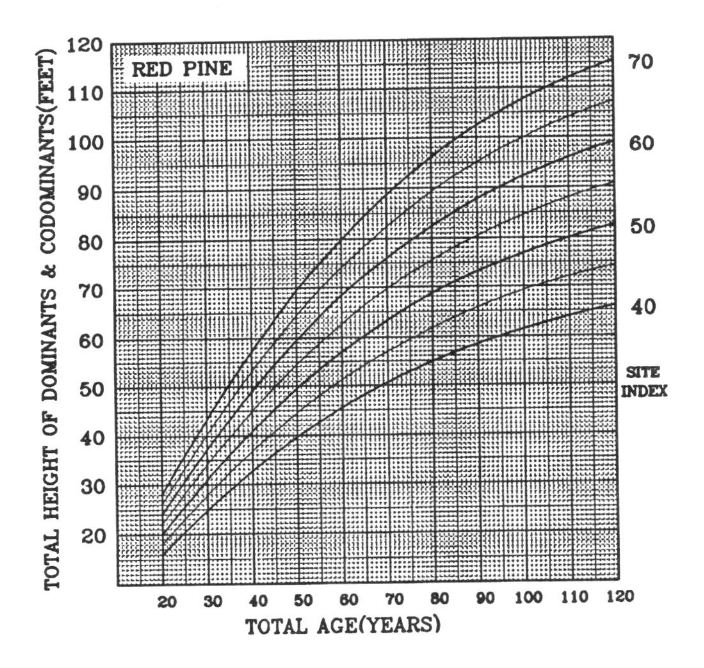

Fig. 17. Site index curves for red (Norway) pine plantations.

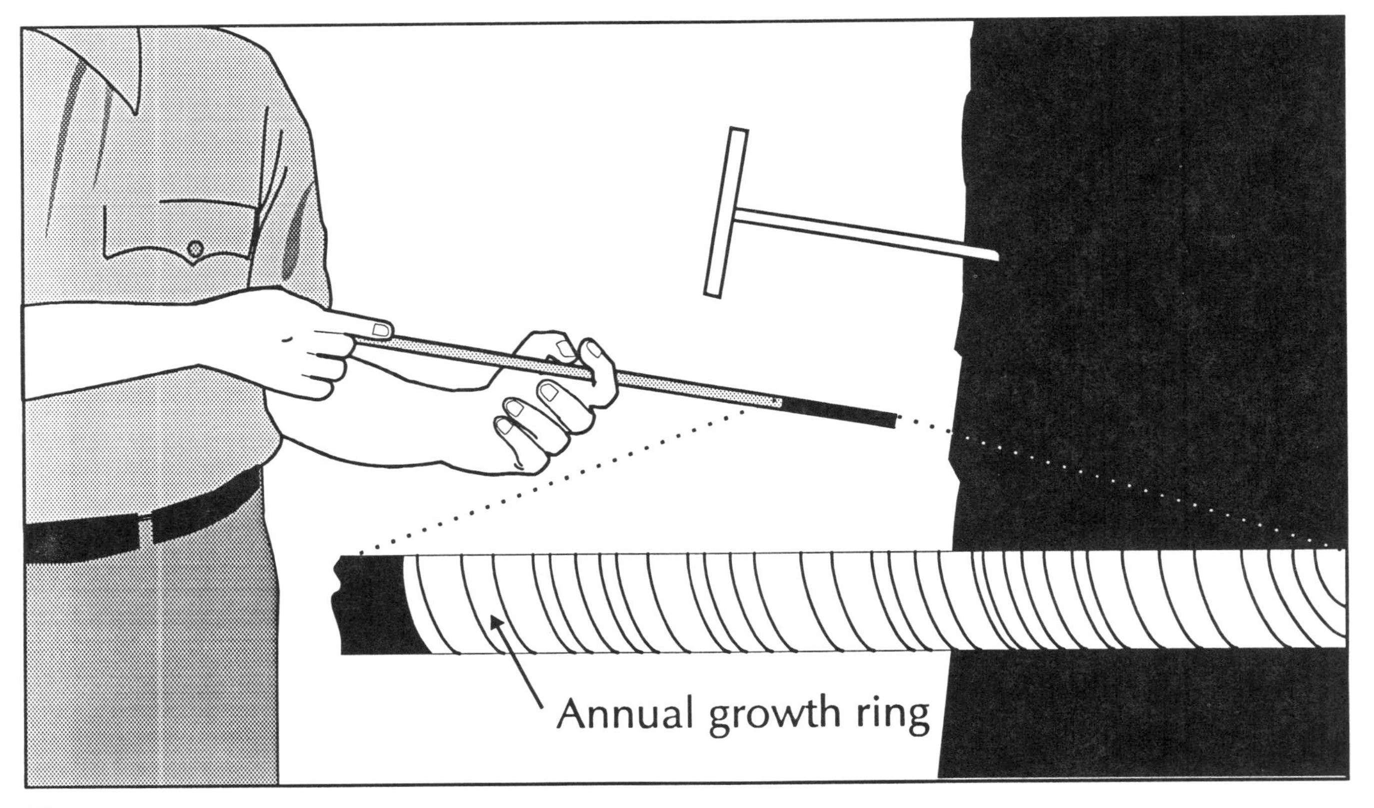

Fig. 16. Tree core sample removed by an increment borer.

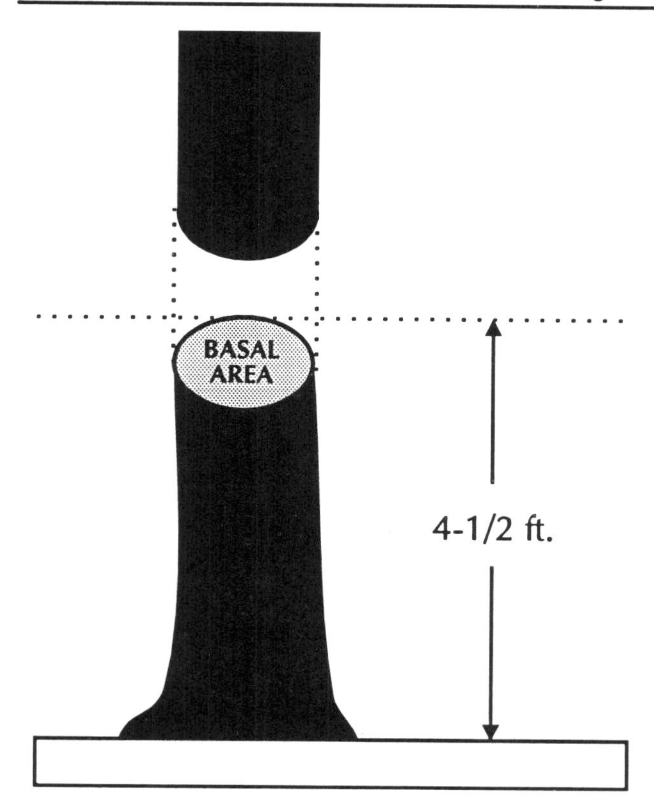

Fig. 18. The basal area of a tree is its cross-sectional area in square feet measured 4-1/2 feet above ground.

of basal area per acre. Be sure to measure BA per acre at several points and calculate an average.

In mature timber stands, BA commonly ranges from 70 to 140 square feet per acre. Dense conifer stands occasionally exceed 200 square feet of BA per acre.

CROWN COVER

The percentage of crown cover is another measure of stand density. This is an estimate of the percentage of ground area in the stand covered by tree crowns (Figure 20). It is a useful indicator of the regeneration potential for certain tree species, given their shade tolerance.

A crown cover of 100 percent is quite dense. Only very shade-tolerant tree species will grow below such a canopy. A crown cover of 40 percent is low. Many different tree species can reproduce and survive in the sunlight that reaches the ground from this low-density crown cover. Guidelines for timber harvests or thinnings sometimes specify the crown cover percentage that should remain after the harvest to stimulate reproduction.

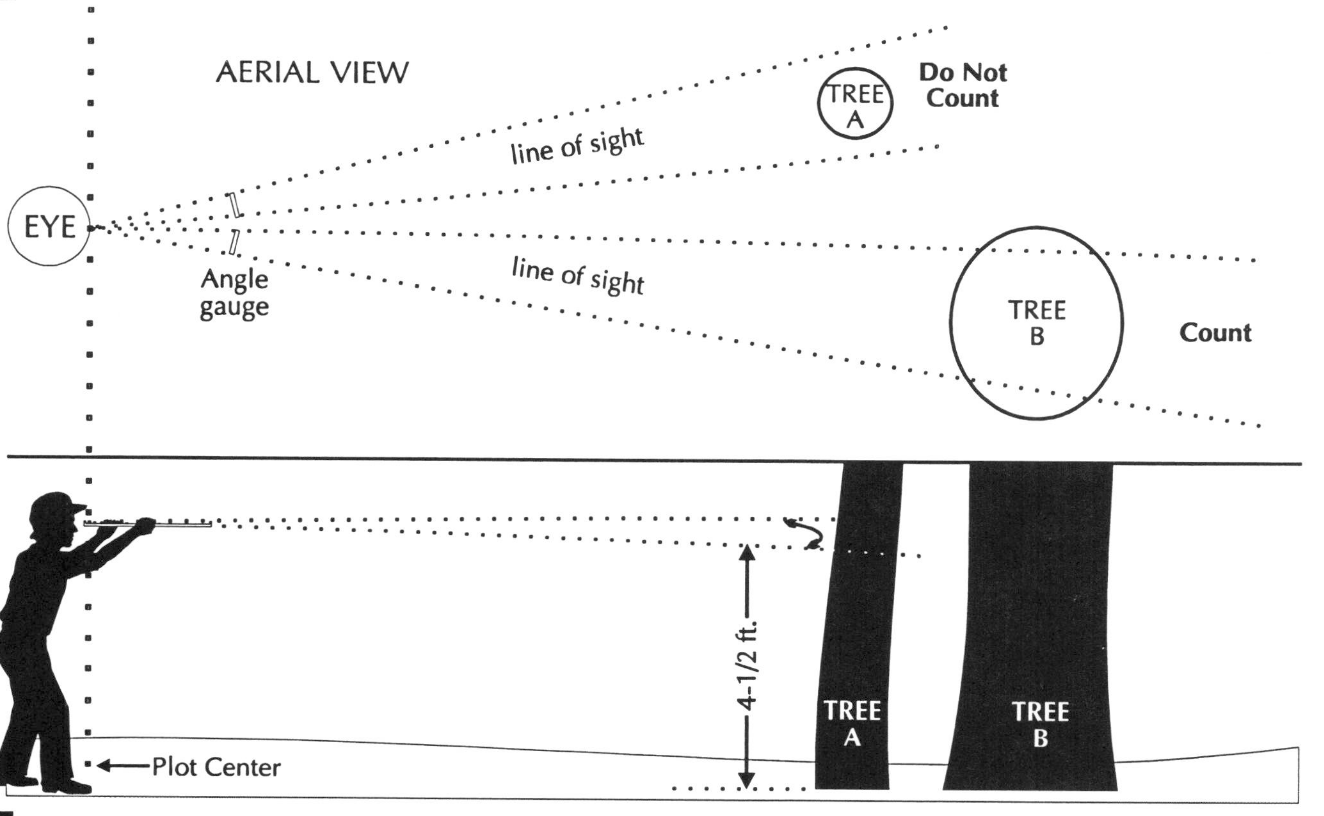

Fig. 19. An angle gauge can be used to measure basal area.

GROWTH AND YIELD

Yield tables have been developed for some commercially important tree species showing the estimated volume of wood per acre that could be grown at different stand ages. Such tables usually estimate expected volumes in pure, even-aged stands on sites of different quality as measured by site index. Table 4 is a yield table for red pine. If you owned an even-aged red pine stand with a site index of 65, current stand age of 80 years, and basal area of 120 square feet, you could expect to have a volume of approximately 42.3 cords per acre. There also are computer programs that predict growth and yield at different stand ages based on inventory information.

Yield information can be used as a basis for financial analyses of different investment alternatives. Yield tables are not always accurate, but are helpful for comparisons. A forester may be able to provide growth and yield information for your woodland.

Suggested References

Avery, T. E. and H.E. Burkhart. 1983. *Forest Measurements.* McGraw-Hill Book Company, New York. 331 pp.

Blinn, C. R. and T. E. Burk. 1986. *Sampling and Measuring Timber in the Private Woodland (NR-FO-3025).* Minnesota Extension Service, Room 3 Coffey Hall, University of Minnesota, St. Paul, MN 55108. 8 pp.

Carmean, W. H., J. T. Hahn, and R. D. Jacobs. 1989. *Site Index Curves for Forest Tree Species in the Eastern United States (General Technical Report NC-128).* USDA Forest Service, North Central Forest Experiment Station, St. Paul, MN 55108. 142 pp.

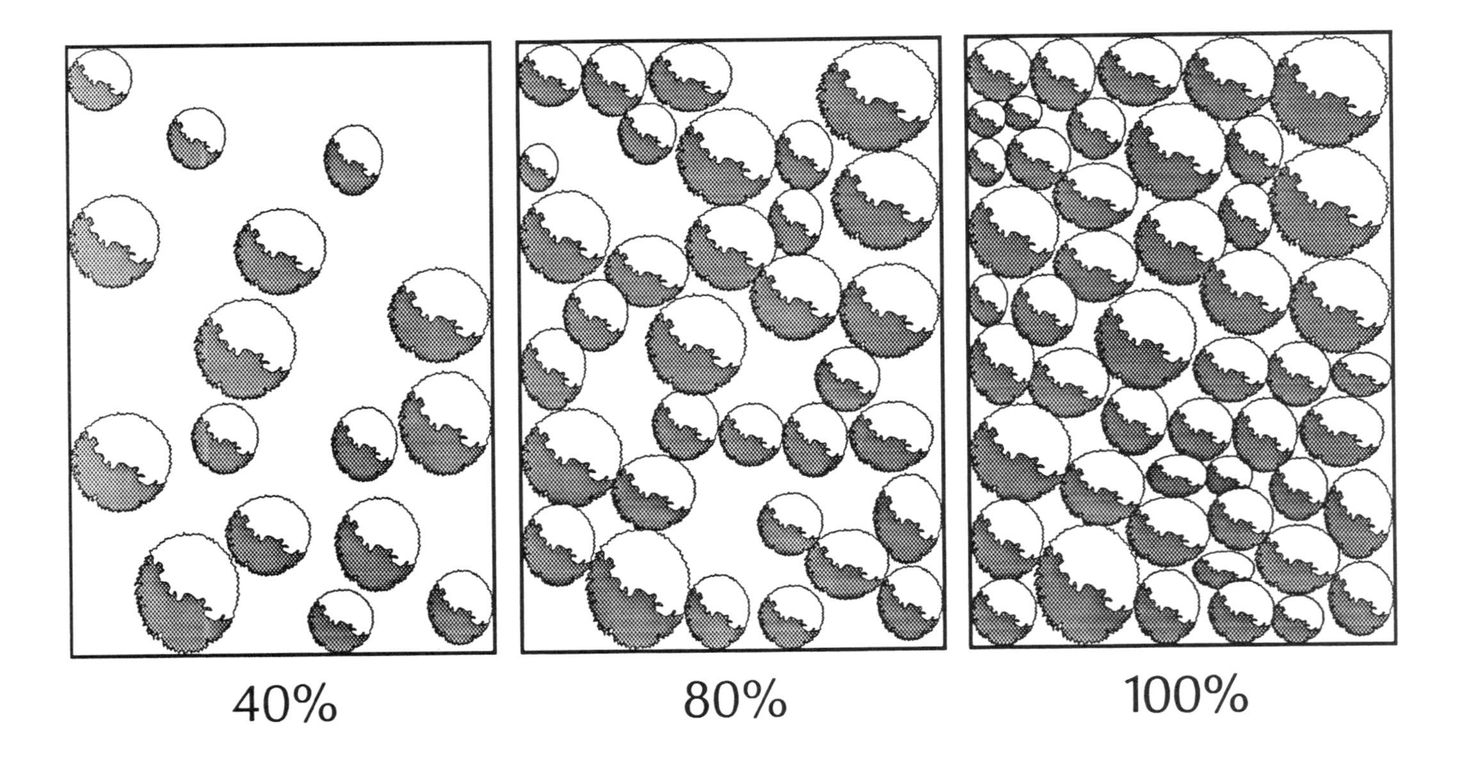

Fig. 20. Percent crown cover.

Table 4. Volume in cords per acre[1] for even-aged red pine stands by site index, age, total tree height, and basal area.

TOTAL AGE IN YEARS	TOTAL HEIGHT IN FEET	BASAL AREA PER ACRE[2]					
		30	60	90	120	150	180
SITE INDEX 75		CORDS PER ACRE					
40	61	7.2	14.5	21.7	29.0	36.2	43.5
60	86	10.2	20.4	30.6	40.8	51.0	61.3
80	103	12.2	24.5	36.7	48.9	61.2	73.4
100	115	13.6	27.3	41.0	54.6	68.3	81.9
120	124	14.7	29.4	44.2	58.9	73.6	88.3
140	130	15.4	30.9	46.3	61.7	77.2	92.6
160	134	15.9	31.8	47.7	63.6	79.6	95.5
SITE INDEX 65							
40	53	6.3	12.6	18.9	25.2	31.5	37.8
60	74	8.8	17.6	26.4	35.1	43.9	52.7
80	89	10.6	21.1	31.7	42.3	52.8	63.4
100	100	11.9	23.7	35.6	47.5	59.4	71.2
120	107	12.7	25.4	38.1	50.8	63.5	76.2
140	112	13.3	26.6	39.9	53.2	66.5	79.8
160	116	13.8	27.5	41.3	55.1	68.9	82.6
SITE INDEX 55							
40	45	5.3	10.7	16.0	21.4	26.7	32.0
60	63	7.5	15.0	22.4	29.9	37.4	44.9
80	76	9.0	18.0	27.1	36.1	45.1	54.1
100	85	10.1	20.2	30.3	40.4	50.5	60.5
120	91	10.8	21.6	32.4	43.2	54.0	64.8
140	95	11.3	22.6	33.8	45.1	56.4	67.7
160	98	11.6	23.3	34.9	46.5	58.2	69.8
SITE INDEX 45							
40	37	4.4	8.8	13.2	17.6	22.0	26.4
60	51	6.1	12.1	18.2	24.2	30.3	36.3
80	62	7.4	14.7	22.1	29.4	36.8	44.2
100	69	8.2	16.4	24.6	32.8	41.0	49.2
120	74	8.8	17.6	26.4	35.1	43.9	52.7
140	78	9.3	18.5	27.8	37.0	46.3	55.6
160	80	9.5	19.0	28.5	38.0	47.5	57.0

[1]Cords = 0.003958 (BA x Height). Rough cords for trees 3.6 inches DBH and larger to a 3-inch top DIB.

[2]For trees 3.6 inches DBH and larger.

SOURCE: Benzie, J. W. 1977. *Manager's Handbook for Red Pine in the North Central States (General Technical Report NC-33).* USDA Forest Service, North Central Forest Experiment Station, St. Paul, MN 55108. 22 pp.

3

HOW TREES GROW

To survive and grow, trees need adequate amounts of carbon dioxide, water, nutrients, and sunlight. Nutrient elements needed in greatest abundance include carbon, oxygen, hydrogen, nitrogen, potassium, calcium, magnesium, phosphorus, and sulfur. Seven other elements are needed in trace amounts to ensure proper growth. Factors that influence the availability or use of carbon dioxide, water, nutrients, and sunlight include site characteristics, climate, and individual tree characteristics.

EFFECT OF SITE CHARACTERISTICS

Site characteristics that affect tree growth include soil depth, texture, moisture, and fertility, and topography.

On the whole, deep soils are better for tree growth than shallow soils because they potentially have a greater nutrient supply and water-holding capacity. Rooting depth may be restricted by bedrock, coarse gravel, a hardpan layer, or excess soil moisture. The tree roots that absorb most nutrients and water are found in the top 2 feet of the soil profile. Some tree species have tap roots. These grow much deeper and serve primarily to anchor the tree, but also absorb water and nutrients.

Soil texture refers to the size of soil particles. Particles are classified by size as sand, silt, and clay. Clay particles are very small. Silt particles are moderate in size. Sand particles are relatively large.

Different soils have different proportions of each particle size. Soils with a high percentage of sand have large pore spaces between soil particles. They absorb water quickly, but water also drains through them quickly, so they tend to be droughty unless there is a shallow water table. Clay soils have a large water-holding capacity, but absorb water slowly and water adheres so tightly to the soil particles that much of it is unavailable for plant use. Soils with a high percentage of silt have the most favorable texture for moisture absorption and drainage.

Soil fertility is based largely on the type of parent material from which the soil originated. Some of the most fertile soils originated from limestone, shale, and windblown deposits, whereas some of the least fertile soils originated from sandstone and granite. On the whole, fine-textured (clay) and medium-textured (silt) soils have a greater nutrient supply than coarse-textured (sandy) soils.

Topography affects tree growth largely because of its influence on soil depth and moisture availability. Because gravity pulls soil particles and water downhill, soil depth, nutrient supply, and water supply usually are greater on bottomlands, lower slopes, and benches than on steep slopes and ridge tops.

Aspect, the direction a slope faces, also influences the amount of sunlight and subsequently soil moisture available to trees. Slopes that face north and east tend to be cooler and moister than slopes that face south and west. This occurs because the sun generally is slightly to the south here in the northern hemisphere and shines more directly on south slopes, warming them and evaporating more moisture. West slopes are drier than east slopes because the sun shines on west slopes during the hottest part of the day, increasing water use by trees and evaporation from the soil. These effects become exaggerated as the steepness of the slope increases.

Too much water is a problem on some sites—for instance, on bottomlands with a high water table or on slopes where bedrock forces water to the surface. Saturated soil causes poor soil aeration, interfering with absorption of water and nutrients by roots and with biological activity in the soil that is necessary for decomposition of organic matter and the release of nutrients back into the soil. Since only a few tree species can grow roots below the water table, a high water table will force most tree species to develop a shallow root system. Table 5 suggests which tree species should be favored for different soil texture and moisture conditions.

Table 5. Tree species suited for different soil texture and moisture conditions.

SOIL MOISTURE CONDITION	SOIL TEXTURE		
	SANDY[1] (COARSE)	LOAMY (MEDIUM)	CLAYEY[2] (FINE)
WET[3]	willow cottonwood	willow cottonwood	willow cottonwood
MOIST[4]	red pine white pine black spruce white spruce	black walnut cottonwood silver maple white spruce white pine	cottonwood green ash silver maple white spruce
MODERATELY DRY[5]	red pine jack pine scotch pine	green ash red oak white spruce	green ash silver maple cottonwood eastern redcedar
DRY[6]	jack pine scotch pine	green ash eastern redcedar	green ash eastern redcedar

[1] At least two-thirds sand.

[2] At least one-third clay.

[3] Subject to standing water from a few hours to a few weeks.

[4] Moist sites in forested areas exclusive of bogs and other sites classified as wet. Includes north-facing slopes with deep soils and areas where water tables are between 3 and 8 feet below the surface.

[5] Nonforested areas in general, exclusive of river bottoms, and level areas where water tables are below 8 feet in drier forested areas.

[6] Water table below 12 feet in nonforested areas, driest sites in forested areas such as southwest-facing slopes, and soils less than 3 feet deep.

SOURCE: Smith, M. E. and H. Scholten. 1981. *Planting Trees in Minnesota (NR-FO-0481).* Minnesota Extension Service, Room 3 Coffey Hall, University of Minnesota, St. Paul, MN 55108. 12 pp.

EFFECT OF CLIMATE

Length of the frost-free growing season, cold temperature extremes, precipitation amount, and duration of droughts are some of the elements of climate that influence tree growth.

Native trees have evolved in our climate and are adapted to it. When we attempt to import trees from the south, they often cannot survive the winters, are damaged by late spring or early fall frosts, or find the growing season too short to consistently produce viable seed. Trees moved north of their original location actually may grow faster than local trees. Trees growing on the northern limits of their natural range, however, usually should not be moved more than 50 miles north. On the other hand, northern species moved south often are unable to survive or compete with native species because of the higher temperatures and greater water demands there. Figure 21 shows plant hardiness zones in the Midwest. Trees and tree seeds usually should not be moved from one zone to another.

In prairie regions, rainfall is not evenly distributed over the growing season and prolonged summer droughts combined with factors such as wildfire limit tree survival and growth. Some tree species will grow in droughty prairie regions, but may need supplemental watering, mulching, or weed control, especially when young.

AVERAGE ANNUAL MINIMUM TEMPERATURE
(in degrees Fahrenheit)

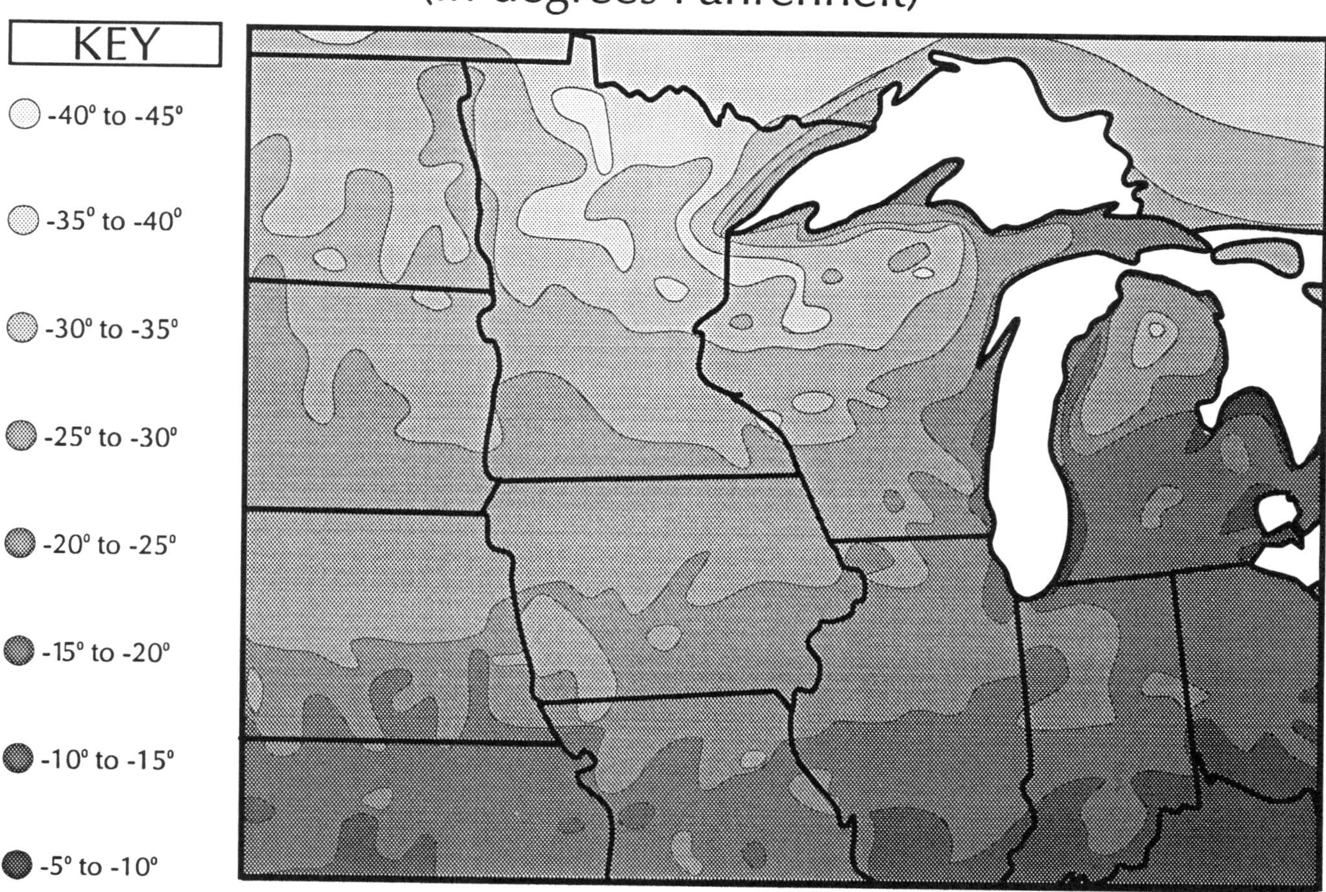

Fig. 21. Plant hardiness zones in the Midwest. (From U.S. Department of Agriculture.)

EFFECT OF TREE CHARACTERISTICS

A tree's crown size, genetics, and ability to tolerate shade and competition from other plants all influence how well it will grow in different environments.

Trees need carbon dioxide, sunlight, water, and nutrients to grow. In general, carbon dioxide and sunlight are absorbed by the leaves, water is absorbed by the roots, and nutrients are absorbed mostly by the roots but also by the leaves.

Leaves are the food factories in a tree. They contain a green substance called chlorophyll that enables the sun's energy to convert carbon dioxide and water into sugar. Trees use this sugar as a basic ingredient to produce wood, leaves, seeds, and other plant parts. Trees with large crowns have more leaves and therefore normally grow faster than trees with small crowns.

The live-crown ratio of a tree is the percentage of total tree height that has live branches on it (Figure 22). For timber production purposes a live-crown ratio of approximately 30 percent usually is optimum, but this varies by species. If the crown is too small, the tree will grow slowly. If the crown is too large, the amount of usable wood in the main stem will be reduced.

Genetics may influence characteristics such as rate of height growth, stem form, self-pruning tendency, and tolerance to insects and diseases. Plant genetically improved stock when it is available, and over time manipulate the composition of your existing stands to favor high-quality trees and eliminate poor-quality ones. If you reproduce your stand naturally, be sure to kill or harvest undesirable trees and permit only high-quality trees to produce seed, stump sprouts, or root suckers. (Poor-quality trees may be an acceptable seed source, however, if their rough appearance is a result of stand conditions or damage not related to genetic characteristics.)

Tree species differ with respect to their tolerance for shade and competition (Table 6). Trees often are classified as: very tolerant, tolerant, intermediate, intolerant, and very intolerant to shade. Those that are very tolerant will reproduce and grow beneath a dense canopy. Trees that are very intolerant will survive only if their seed sprouts in openings that receive direct sunlight. You must know the shade tolerance of a species to determine the site conditions necessary for reproducing it. The shade tolerance of a tree may change somewhat as it grows older.

Fig. 22. Live-crown ratio of a tree.

Table 6. Approximate shade tolerance of selected tree species.

Very Tolerant	Tolerant	Intermediate	Intolerant	Very Intolerant
balsam fir	American basswood	American elm	black ash	black willow
eastern hemlock	black spruce	bitternut hickory	black cherry	eastern cottonwood
ironwood	northern white-cedar	eastern white pine	black walnut	eastern redcedar
sugar maple	white spruce	green ash	butternut	jack pine
		northern red oak	paper birch	quaking aspen
		red maple	red pine	tamarack
		shagbark hickory	silver maple	
		sycamore		
		white ash		
		white oak		
		yellow birch		

ADAPTED FROM: U.S. Department of Agriculture, Forest Service. 1990. *Silvics of North America, Volume 1 Conifers (Agricultural Handbook No. 654).* U.S. Government Printing Office, Washington, DC 20402. 675 pp.

U.S. Department of Agriculture, Forest Service. 1990. *Silvics of North America, Volume 2 Hardwoods (Agricultural Handbook No. 654).* U.S. Government Printing Office, Washington, DC 20402. 877 pp.

Miller H. and S. Lamb. 1985. *Oaks of North America.* Naturegraph Publishers, P.O. Box 1075, Happy Camp, CA 96039. p. 72.

4

REGENERATING WOODLAND STANDS

A stand should be regenerated when most of its trees are economically mature, when it is stocked with undesirable tree species or poor quality trees, or when it is greatly understocked. Tree stands may be reproduced through natural or artificial means or by a combination of the two. The choice of how best to reproduce a stand depends on the tree species involved, financial considerations, and management objectives.

NATURAL REGENERATION

Depending on the species, trees may reproduce naturally by seeds, root suckers, stump sprouts, or layering (Figure 23).

Seeds

When regenerating stands by natural seeding, it is important that you know how the species you wish to encourage disperses seed, how far its seed travels, how abundantly it produces seed, and what type of seedbed is needed for germination. This information will affect how you harvest and prepare the site.

Tree seeds are dispersed in a variety of ways. Aspen and cottonwood seeds are covered with a cottonlike down and may be carried several miles by the wind. Maple and pine seeds have wings allowing them to glide in the wind. Cherry seeds frequently are dispersed by birds that eat the cherries and drop the seeds far from parent trees. Walnuts, acorns, and pinecones often are carried away and buried by squirrels. Seeds from willows and other shoreline species may be dispersed by water.

Very few of the seeds produced grow into seedlings. A number of conditions must be met for natural seeding to be successful. In addition to an adequate supply of viable seed, there must be a receptive seedbed. For most species, this means that mineral soil must be exposed so the seed can get enough moisture to germinate and grow. Soil temperatures must be high enough so seeds will germinate, but not so high that the seedlings will be killed. Rodent predation of seed, available moisture, and vegetative competition also affect success or failure.

Most hardwood species reproduce easily from seed when conditions are favorable. However, natural reproduction of conifers from seed often is erratic and unpredictable. Jack pine and black spruce can successfully regenerate from seed when seedbed, moisture, and temperature conditions are right. These two species often store seed in cones on the tree for many years until heat, as from a fire, opens the cones and releases the seed. For other conifers such as red (Norway) pine, white pine, white spruce, and tamarack, viable

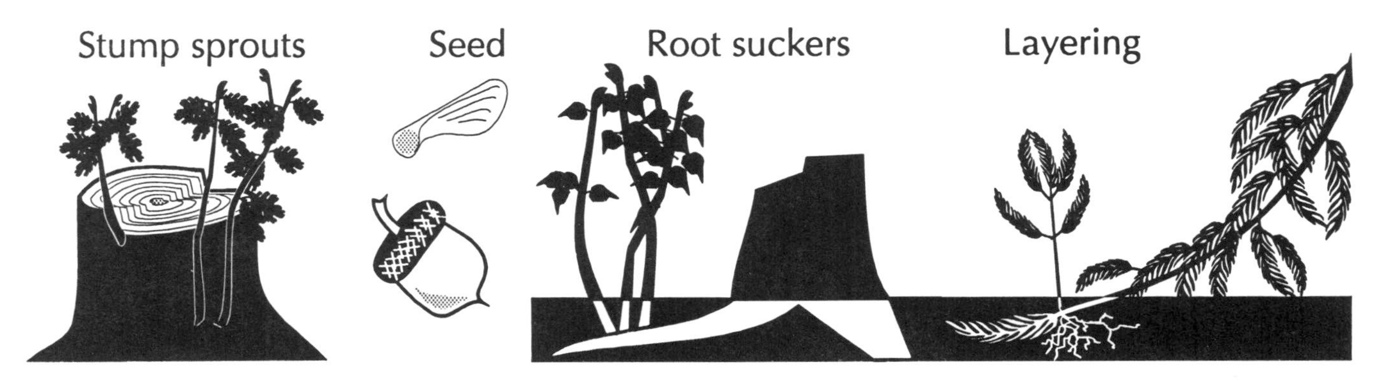

Fig. 23. Natural regeneration methods.

seed crops are unpredictable and successful stand establishment does not always occur.

Root Suckers

Some hardwoods (e.g., aspen and black locust) regenerate from root suckers as well as from seed. A tree that grows from a root sucker is genetically identical to the parent tree and is called a clone. Suckers usually develop after a parent tree has been cut down. A single parent tree may produce several hundred suckers, creating a dense new stand. The number of suckers may be reduced if the parent tree is in poor health or if timber harvesting causes root system damage or soil compaction.

Stump Sprouts

Other hardwoods, including oak, basswood, birch, and maple, sprout from stumps as well as grow from seeds. Relatively young stumps, cut close to the ground in late fall or winter when there are food reserves stored in the roots, sprout the best. Stumps often send up numerous sprouts, but these usually thin naturally to two or three main stems. You can speed up this process and encourage the strongest stump sprouts by cutting the others when the sprouts are five to ten years old.

Layering

Layering occurs when a buried branch takes root and develops into a new tree. The lower limbs of black spruce, balsam fir, and northern white-cedar sometimes touch the ground and become covered with organic matter. Layering is not an important reproduction method in forests, but can provide additional northern white-cedar for deer browse on suitable sites.

ARTIFICIAL REGENERATION

Artificial regeneration includes direct seeding, planting seedlings, and planting cuttings. It usually is more expensive than natural regeneration, but permits better control over species selection, genetic characteristics, and tree spacing.

Direct Seeding

Direct seeding is the process of sowing or planting seeds. It often is used to establish jack pine and black spruce, as well as some hardwoods, including black walnut. It is relatively inexpensive, but generally is less reliable than planting seedlings. Direct seeding of black spruce is preferred to planting seedlings on sites with poor access, such as spruce bogs. The appropriate site preparation, moisture, and temperature requirements vary by species and are similar to those necessary for natural seeding. Often the seed is chemically treated to protect it from diseases, rodents, and birds.

Seedlings

Planting seedlings, either bare-root or container-grown stock, is the most reliable way to regenerate a stand, especially for conifers. Bare-root seedlings are the most common. They frequently are designated as 1-0, 2-0 or 2-1 stock, with the first number referring to how many years they were grown in the original nursery seedbed and the second to how many years they were grown in a transplant bed.

Transplants—seedlings that spent a year in a transplant bed—generally have a more fibrous root system and larger stem diameter than seedlings that were not transplanted. Transplants are recommended for regenerating slow-growing conifer species such as spruce and fir, and for harsh planting sites where survival is likely to be a problem.

Seedling costs vary depending on tree age, grade, species, and quantity ordered. Transplants survive very well, but are expensive and therefore are not widely used. One- or two-year-old seedlings are less expensive than transplants and are recommended for most hardwood and conifer plantings. Tree seedlings sometimes are graded and sold by height class or stem diameter.

Container-grown seedlings usually are grown in a greenhouse in containers between 1 and 2 inches in diameter. Some biodegradable containers may be planted in the ground with the seedling in them. Others must be removed from the seedling before it is planted. Container-grown stock can be very useful for dry planting sites or for planting late in the growing season.

Cuttings

Cuttings provide another alternative for artificially regenerating certain tree species. These usually are 8- to 12-inch lengths of tree stems about 1/4- to 3/4-inch in diameter (longer cuttings may be used on drier sites). They are cut during the dormant season from the previous year's growth of vigorous seedlings or stump sprouts. Cuttings usually have no visible roots, but when buried vertically with just an inch of the stem protruding above ground, they will form roots. Rooted cuttings also may be available for purchase.

Cuttings produce an exact genetic replica of the parent tree. They commonly are used to regenerate poplars, but also can be used to regenerate willow and green ash. Cuttings grow best where the soil remains moist throughout the growing season.

Tree Spacing

When designing a plantation, you need to determine an appropriate spacing between trees. Consider the crown width of trees when they reach a useful size. For example, when growing trees for timber, allocate space so individual trees are just beginning to crowd each other when the trees are large enough to support a commercial thinning. A forester can help you determine the correct spacing depending on the species and purpose for the plantation.

Table 7 shows the number of trees needed per acre for various spacings. To calculate the number of trees per acre for other spacings, multiply the planned spacing (in feet) within rows by the spacing (in feet) between rows and divide that number into 43,560. For example, if trees are to be spaced 8 feet apart within rows and rows are to be 10 feet apart, you would plant 545 trees per acre:

$$\frac{43{,}560}{8 \times 10} = 545 \textit{ TREES PER ACRE}$$

Table 7. Number of trees per acre at different spacings.

SPACING (IN FEET)	TREES PER ACRE
4 x 4	2,722
5 x 5	1,742
6 x 6	1,210
7 x 7	890
8 x 8	680
9 x 9	538
10 x 10	436
11 x 11	368
12 x 12	303

Site Preparation

Site preparation often is necessary prior to planting. Its purposes usually are to expose mineral soil and set back competing vegetation. While site preparation can be done with hand tools, this method usually is expensive and vegetation that is cut down may resprout. There are herbicides available for site preparation that may be very effective and economical in some situations. You also can use mechanical methods such as disking, scalping, or trenching. If you use mechanical methods, be careful not to destroy the nutrient-rich organic layer near the soil surface. Often a combination of mechanical and chemical methods is most effective. In some circumstances controlled burning can be used to remove debris from sites and temporarily reduce vegetative competition.

Planting

The best time to plant is in the spring as soon as the frost is out of the ground. At this time the soil is moist, the climate is somewhat mild, and normally there is ample rainfall. If necessary, container-grown seedlings can be planted later in the growing season. Fall planting usually is less successful because frost heaving may occur, especially on fine-textured or wet soils, and growth regulators in the tree may become imbalanced, leading to top dieback.

Be sure to take good care of seedlings before you plant them. They should be dormant when you receive them (i.e., the buds should not be elongating or flushed).

It is important that you not let the roots dry out or freeze. This will likely kill the trees, but since the crown will appear alive (conifers will remain green) for a period of time, there is no obvious clue that the trees are worthless. To minimize risk of tree roots drying during shipment, ask nurseries to ship by the swiftest transportation available. If you transport the trees yourself, protect them from wind and sun during transit.

The Minnesota Department of Natural Resources packages its nursery trees in a plastic bag inside a wax-lined cardboard box. It recommends the following handling procedures:

1. Protect the box of seedlings from direct sunlight and heat.
2. Plant the seedlings as soon as possible. Seedlings should not be stored for more than three to five days and then only at temperatures of 35^0 to 45^0F.
3. Once you open the package, plant the seedlings immediately. Exposure of tree roots to hot sunlight and drying winds for three to five minutes may be fatal.
4. While planting, keep the seedling roots moist, but do not immerse them in water for more than an hour.
5. If you need to postpone planting for more than three to five days after receiving the trees, remove the seedlings from the container and heel them in a trench.

To heel-in trees, dig a V-shaped trench in a cool, shady location, deep enough so the earth will

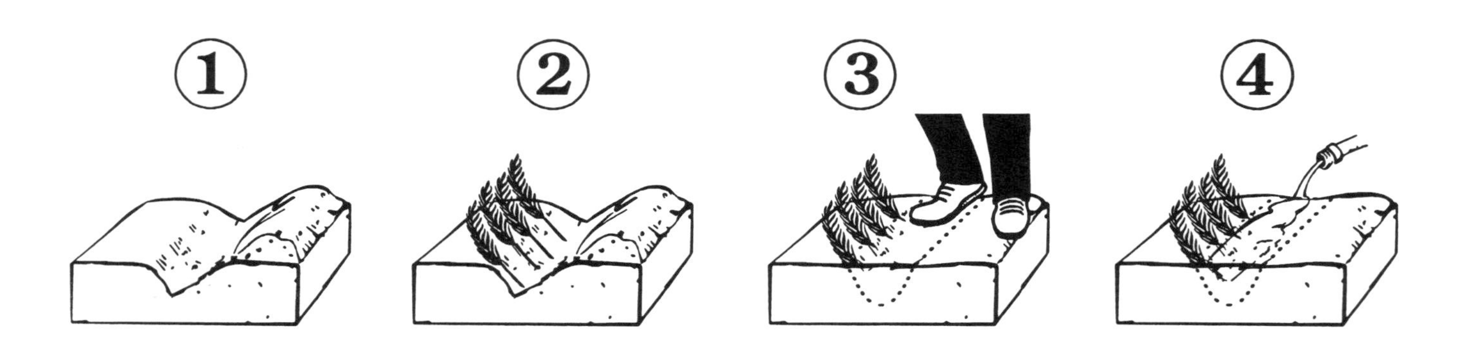

Fig. 24. Heel-in seedlings for temporary storage.

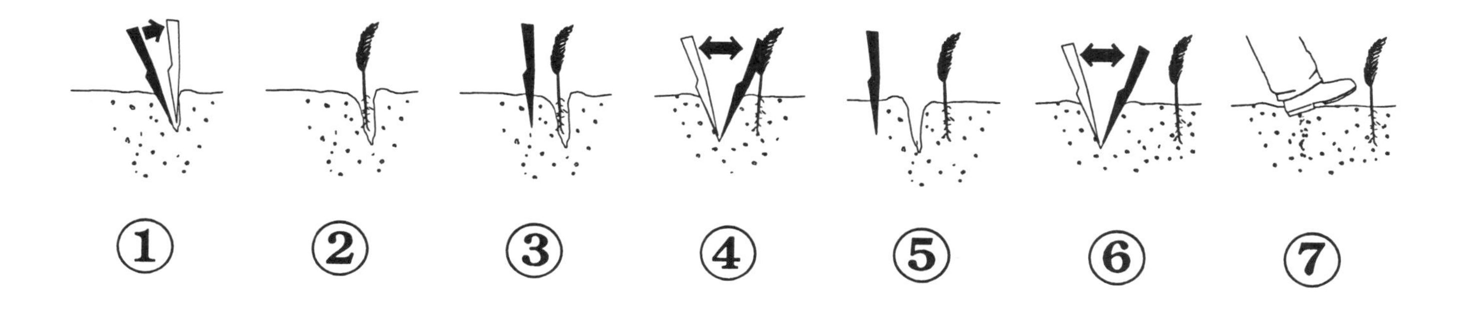

Fig. 25. Planting seedlings using the slit or bar method.

cover the entire root system and part of the lower stem (Figure 24). Open the boxes, spread the trees along the sloping side of the trench in two or three layers, pack soil around the roots, and water as necessary to keep the roots moist. Store trees in this manner only as long as they remain dormant.

An even better storage method is to place the seedlings in a refrigerator set at 40^{0}F or slightly cooler.

It is absolutely essential that seedlings be planted before new growth starts to emerge. The sooner you plant the trees after they arrive from the nursery, the better the survival will be.

You may plant trees by hand or machine. Regardless of the method, follow these rules:

1. Plant the tree at the same depth that it grew in the nursery.
2. Plant the tree in a vertical, upright position to avoid a crooked stem.
3. Place the roots in the planting hole in a normal position without twisting or bending.
4. Carefully firm the soil around the roots to eliminate air pockets.
5. Plant only when soil moisture is adequate to ensure survival.

There are two general methods of hand planting. One of these is the hole method. Dig a hole with a shovel, mattock, or grub hoe. It should be large enough to accommodate the tree roots without bending. Place the tree in the hole, distribute the roots evenly, and pack the soil firmly around the roots, covering the root collar. This method usually results in a high rate of survival, but it is slow and is not practical for planting large numbers of trees.

The slit or bar method (Figure 25) is preferred when a large number of trees are being planted because it is faster. Insert a spade, planting bar, hoedad or similar tool into the soil and move it back and forth to form a V-shaped slit. Insert the tree seedling into the slit so it will be buried to the root collar or to the same depth the tree was growing in the nursery. (If you err at all, plant slightly too deep rather than too shallow.) Remove the planting bar and reinsert it about three inches behind the seedling. Pull the bar back to firm soil around the roots, then push forward on the bar to seal the top of the planting hole. Push soil into the second slit and press down firmly with your boot to seal the slit. Using this method, you can plant 1,000 to 3,000 seedlings per day, depending on your experience and the condition of the planting site.

There are many designs for tree-planting machines, but generally they have a coulter that breaks through the soil surface, a V-shaped blade

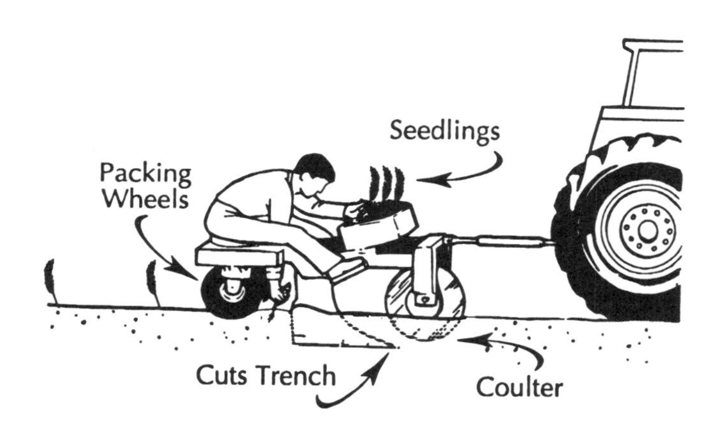

Fig. 26. A tree-planting machine.

that opens a trench into which the operator places seedlings, and packing wheels that firm the soil around the seedlings (Figure 26). Some newer planting machines have spray attachments for applying herbicides for grass and weed control.

Tree-planting machines work best where terrain is fairly level and the site has been cleared of stumps and logging debris. If you use a tree-planting machine, protect seedlings from the wind and sun so the roots do not dry out. Place seedlings erect in the planting trench at the proper depth and pack the soil firmly around the roots. A three-person crew using a tree-planting machine can plant about 10,000 trees in an eight-hour day.

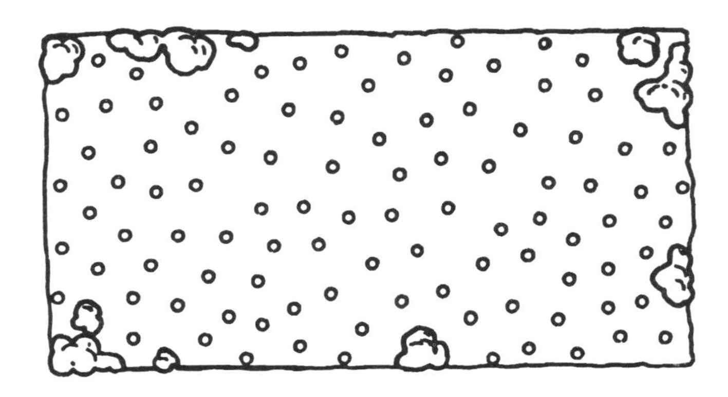

Fig. 27. Clearcutting system.

SILVICULTURAL SYSTEMS

A silvicultural system (sometimes called a regeneration system) is a combination of timber harvesting and site preparation practices that prepare a site for natural or artificial regeneration. The four common silvicultural systems are clearcutting, seed-tree, shelterwood, and selection. The main differences among these methods are the age distributions of the regenerated stands and the degree of exposure of the forest floor after a harvest. Which system you use depends on growing conditions needed by the species you want to favor, economic considerations, whether you want to create an even-aged or uneven-aged stand, and management objectives (timber, wildlife, aesthetics). Further information concerning application of these regeneration systems to particular forest types is given in Chapter 6.

Clearcutting System

There are two general types of clearcutting (Figure 27). Complete clearcutting involves harvesting every tree in the stand regardless of species, size or marketability. This practice usually creates the best site conditions for regenerating the stand. Commercial clearcutting involves harvesting only trees of marketable species, size, and quality. Commercial clearcutting, sometimes called "high-grading," is appropriate only where nearly all trees in the stand are marketable or where the harvesting process will knock down all the unmerchantable trees. Otherwise, if only high-quality trees of desirable species are taken, only genetically inferior trees and undesirable species are left to regenerate the stand.

Clearcuts usually are larger than 2 acres, but they vary greatly in size and shape. Sometimes only strips or patches of timber are harvested and the uncut strips or patches are harvested only after the previously harvested areas have regenerated. This method may be more pleasing to the eye than clearcutting large blocks, and sometimes the uncut areas are necessary to provide seed for regenerating the harvested strips or patches.

Clearcutting usually is appropriate to achieve one or more of the following objectives:

- To regenerate trees that need full sunlight for germination and seedling growth.
- To regenerate shallow-rooted species or trees growing in exposed locations where trees left standing after a harvest would be uprooted or broken by wind.
- To produce even-aged stands.
- To regenerate species that naturally reproduce from seeds scattered by the wind, from root suckers, or from seeds released from cones after fires.
- To salvage all merchantable material when whole stands are over-mature and need regenerating or when stands have been killed by insects, disease, wind, or fire.

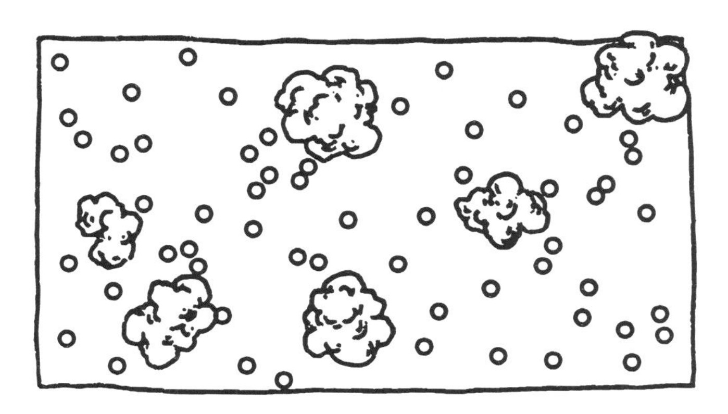

Fig. 28. Seed-tree system.

- To clear the site for conversion to another species by planting or seeding.
- To provide essential habitat for wildlife species that require high density, even-aged stands.

Clearcutting is common with jack, red, and white pine; white and black spruce; aspen; and, to some extent, oak. These species cannot tolerate too much shade, and grow best in even-aged stands.

When clearcutting on slopes and erodible soils, take care to design logging roads and harvest timber in a way that minimizes soil erosion. Clearcuts also should be regenerated as soon as possible after cutting so the area will not regenerate with undesirable species.

Some people object to the temporarily cleared landscape, but there are sound biological and economic reasons for clearcutting. Clearcutting does not ruin the land or timber productivity, if properly done. Many tree species reproduce best when the stand is clearcut.

Seed-Tree System

This system leaves trees scattered throughout a harvested stand at intervals close enough to furnish seed to the entire cut area (Figure 28). Once the new forest is established, the seed trees may be harvested or left to grow through the next rotation. The seed-tree system is appropriate only for tree species that produce wind-dispersed seed. To obtain maximum seed production, seed trees should be healthy, large-crowned, and wind-firm.

Seed trees usually are left singly, but may be left in small groups for wind protection. The number of seed trees required depends on the species' ability to produce seed and the distance the seed can be carried by the wind. In most cases five to ten seed trees per acre are left uncut.

The initial cut should occur after seeds have been dispersed in a year when there is a good seed crop. Some form of site preparation usually is necessary to produce a receptive seed bed (see Chapter 6).

The main disadvantages of the seed-tree system include:

1. Diseases carried by the seed trees could be quickly transferred to the seedlings. If seed trees are diseased, do not use the seed-tree system.
2. Seed trees may be killed or damaged by wind, fire, or insects before they produce seed.
3. It may not be economical to harvest seed trees after regeneration occurs.
4. The harvest of seed trees could damage new seedlings.
5. There is a potential for long delays in seed crops and subsequent invasion of competing brush or undesirable tree species.
6. There is little control over the spacing and stocking rate of reproduction.

The seed-tree system is considered to be more aesthetically pleasing than clearcutting. However, because disadvantages are severe, the seed-tree system rarely is practiced in this region.

Shelterwood System

The shelterwood system (Figure 29) is similar to the seed-tree system, but more trees are left and the new stand is established under the partial canopy of older trees. Then the overstory is removed to release the reproduction. Two harvests commonly are made. Occasionally three harvests are necessary.

In a three-harvest system, the first is a preparatory cut or heavy thinning that leaves the best trees with plenty of growing space to expand their crowns, grow vigorously, and produce seed. This cut can be eliminated if intermediate thinnings have achieved the same results.

The second cut, known as the seed cut, is made several years later when there is a good seed crop. The seed cut allows more sunlight to reach the forest floor, stimulating new seedling growth. During harvesting the site is prepared for seed germination by exposing mineral soil and undesirable trees, shrubs, or herbaceous plants in the understory are controlled. In a few years new seedlings establish themselves beneath the shelter of older trees.

The final cut removes the remaining mature trees, completely releasing the young stand. This cut must be made soon enough to maintain the even-aged characteristics of the new stand or a two-aged stand will develop.

The shelterwood system is used to develop advance reproduction before a final harvest. It is most appropriate where the species to be regenerated can grow under partial shade; where seed trees are not subject to windthrow, wind damage, epicormic branching, or logging damage; and where the increased cost of several partial cuts is acceptable. Since a seedling stand is present when the final cut is made, the shelterwood system is slightly more aesthetic than the clearcut system.

There are several variations of the shelterwood system:

- Uniform method—harvest trees scattered throughout the entire stand.
- Group method—harvest groups of trees.
- Strip method—harvest strips of trees in an alternating or progressive pattern.

The shelterwood system sometimes is used to regenerate stands of spruce, balsam fir, white pine, oak, and northern hardwoods such as white birch. It cannot be used with jack pine or aspen because they are very intolerant of shade. It is not recommended for red pine because of the danger of disease being spread from older seed trees to the new seedlings. The shelterwood system does not provide much control over the spacing and number of trees in the new stand.

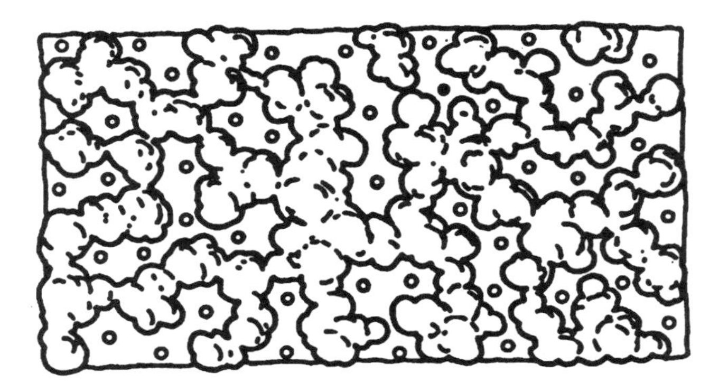

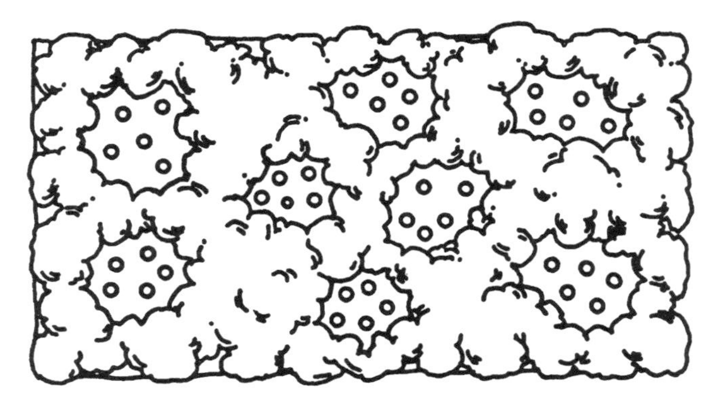

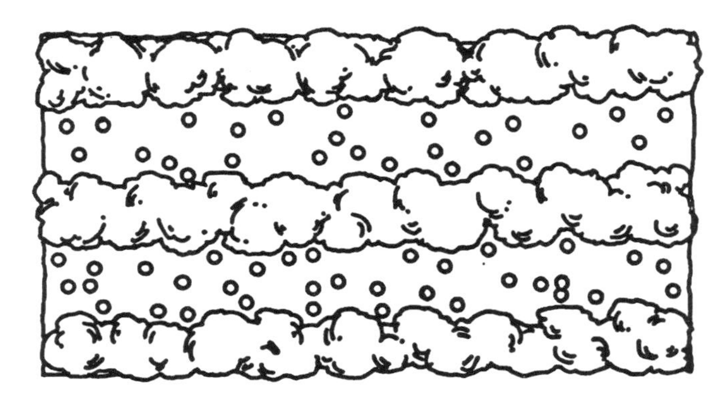

Fig. 29. Shelterwood system.

Selection System

Individual trees or small groups of trees are harvested in the selection system (Figure 30). The cutting interval usually is five or ten years, but may be longer. Trees of all sizes are removed in each cut according to these general guidelines:

1. Harvest trees that are overmature and growing slowly; trees that have poor form, damage, or health; and undesirable species.
2. Aim for proper spacing even if it means leaving a few trees that otherwise should be cut.
3. Make light cuts at least every ten years to encourage regular growth and prevent severe disturbances to the forest.

All too often the selection system is improperly used and timber stands are high-graded. High-grading refers to harvesting only large diameter, high quality trees of merchantable species while leaving trees that are not merchantable because of small size, poor quality, or undesirable species. Inferior trees, often with poor genetic characteristics, then make up the next stand. This is not a desirable practice for any land steward and it is not proper application of the selection system.

The selection system, when properly applied, produces an uneven-aged stand with trees of many ages and sizes present. The average quality of residual trees should improve after each harvest. Generally natural reproduction is expected, but planting is possible in openings left after harvesting mature trees. The single-tree selection system regenerates predominantly shade-tolerant species because of the small canopy openings created. Under the group selection system, larger openings up to a half-acre in size are clearcut, making it possible to regenerate species that are intermediate in shade tolerance.

The selection system can be used to regenerate northern hardwood stands—especially sugar maple, basswood, and ash—as well as spruce-fir stands. It is considered by many to be the most aesthetically pleasing silvicultural system because of the continuous forest cover and small amounts of scattered slash. Major disadvantages of this system are:

1. Each harvest damages some residual trees unless done very carefully.
2. The small volumes of wood harvested per acre and the scattered locations of the trees harvested greatly increase logging costs over other systems.
3. The shade-tolerant tree species that are regenerated may be less desirable than intermediate and intolerant species.

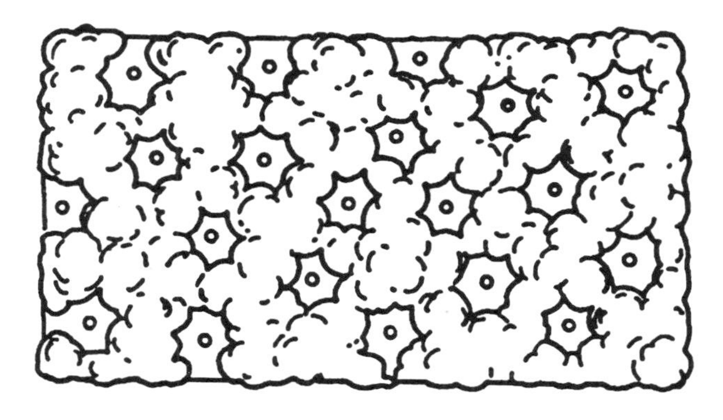

Single-tree selection

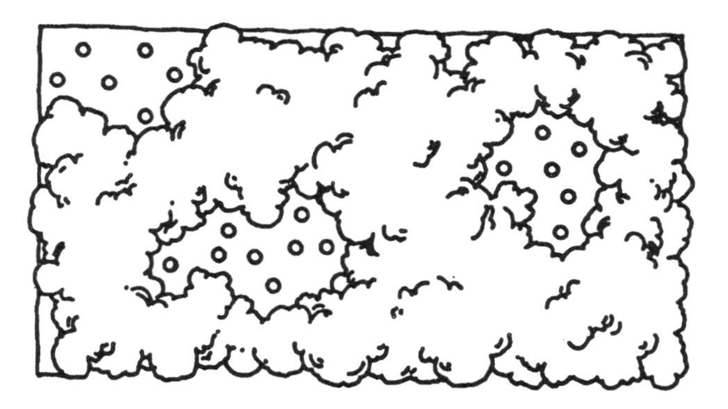

Group selection

Fig. 30. Selection system.

5

WOODLAND IMPROVEMENT PRACTICES

Regenerating a stand is only one aspect of woodland stewardship. Other practices may be necessary to ensure that the most desirable trees survive and grow at optimum rates. These practices will vary depending on the number, size, and species of trees in the stand as well as your management objectives.

Trees often are classified by size as seedlings, saplings, poletimber, or sawtimber. Definitions vary between regions, but the following are widely accepted classifications: a seedling is less than 1 inch DBH, a sapling is 1 to 4 inches DBH, poletimber is 5 to 9 inches DBH, and sawtimber is 10 inches DBH and larger.

SEEDLING AND SAPLING STANDS

When trees are seedling and sapling size, they may need release from herbaceous weeds or woody plants that compete for moisture and sunlight. Release from herbaceous plants and small woody plants can be done manually by weeding and clipping, mechanically by mowing or cultivating, or chemically by using herbicides. Seedlings and saplings also may need liberation—removal or killing of large, undesirable trees that interfere with development of a new stand beneath the canopy.

Hand release is expensive and time-consuming on large areas. It also is inefficient because the weeds and brush continually resprout and grow. When clipped back, some woody plants will send up twice as many new shoots that continue to compete with tree seedlings. Hand release is seldom used on large forested tracts but can work in small plantings.

Mowing and cultivation are appropriate only for tree plantations in old fields where the terrain and lack of woody debris on the ground permit access by vehicles. Both practices will reduce competition from herbaceous plants, but they usually need to be repeated several times during the first three years after the trees have been planted. Cultivation provides better weed control than mowing, but soil erosion is more likely to occur after cultivation. Also, mowing and cultivation equipment sometimes damage the stems, branches, or roots of desirable trees.

The most common method of release is to use herbicides. Usually a single application is sufficient. Herbicides may be applied aerially or from the ground. Ground application using backpack or hand-held sprayers is more common on small plantations.

In order to be used for tree release, a herbicide must be labeled for that use. Read the label carefully and follow instructions for application rates, timing, and precautions to be taken. Herbicide recommendations are constantly changing as products enter and leave the market.

Different herbicides kill vegetation in different ways. Some are growth regulators; others destroy chlorophyll or essential plant tissues. Herbicides must be applied at a time when the target species are most susceptible. For example 2,4-D and glyphosate are absorbed through foliage; application of these herbicides for release of conifers is most effective when the conifer seedlings have set buds but the undesirable vegetation is in full leaf and actively growing. Spraying too early in the growing season before the conifers have set buds may damage them. Other herbicides, such as hexazinone, are applied directly to the soil early in the growing season. Rain moves the chemical into the soil where it is absorbed by the target species' roots.

Use caution when applying herbicides so spray does not drift onto adjacent areas or contaminate surface or ground water. Use drift-control additives if there is any possibility of drift. Soil-activated herbicides can flow downslope and kill nontarget vegetation. Carefully observe environmental protection sections on the label and maintain adequate buffer strips near water and nontarget areas.

A young stand may need liberation when the overtopping canopy is composed of trees with poor quality wood or undesirable species. When these trees have large crowns, they commonly are called wolf trees. Undesirable trees can be harvested if they have any commercial value. Otherwise they can be killed by girdling, frilling, basal spraying, or injecting herbicide as described in the next section on poletimber stands.

POLETIMBER STANDS

As stands grow to poletimber size, they may need weeding, culling, or thinning. Weeding refers to removal of undesirable tree species that take up valuable growing space and that do not contribute to management objectives. Culling is the removal of trees that have no commercial value because of poor form, damage, or other physical defects. Thinning involves selectively cutting trees of a desirable species in order to give residual trees more

Pesticide Applications

Pesticides are any chemicals used to control weeds, insects, diseases, or rodents. They can generate substantial benefits when applied correctly, but can also cause serious damage or personal injury if used incorrectly. Here are some basic application guidelines.

! Apply pesticides only to control pests known to be established in the area.

! If you do not have experience applying pesticides to forests or are not licensed for applying them to forested areas, work with a licensed, professional applicator.

! ALWAYS follow label directions. The label contains valuable information about safety, uses, application rates, mixing, environmental hazards, and things not to spray.

! During ground applications, never spray if the wind speed is over eight miles per hour or if the air temperature is over 85°F. Use low-volatile formulations or drift retardants where indicated.

! AT A MINIMUM, wear protective clothing such as a long-sleeved shirt and pants, rubber gloves, and a hat. **Eye protection is essential during mixing.** Check the label for specific recommendations on personal protection.

! Mix only as much pesticides you need. It is better to underestimate and have to mix a little more than to have leftover material. If you have a little spray mix left over, refill the tank and spray the diluted mix over the treated area. NEVER dump spray mix into a lake, stream, sewer, ditch, or soil pit. Dilute it and use it!

! Triple rinse the container and apply the rinse material to the treated area. Properly dispose of all pesticide containers according to label directions. **Never reuse any pesticide container for any other purpose.**

growing space. In mixed species stands, weeding, culling, and thinning usually are carried out simultaneously. These practices collectively accomplish timber stand improvement or TSI (Figure 31). The objectives of TSI are to:

- Improve species composition.
- Improve stand quality.
- Control stand density.
- Increase tree size.
- Harvest trees before they die.

TSI may be commercial or noncommercial depending on whether harvested trees produce a profit.

Natural timber stands usually have a mixture of tree species, and plantations usually are invaded by tree species not originally planted. Selection of crop trees to be left standing is based on which species or specimens have the greatest value to you. Tree species vary in their wood product, wildlife, and aesthetic values.

If timber production is your goal, favor species that produce wood in high market demand, and healthy specimens that have straight stems. Remove trees with poor form, too many limbs on the main stem, damaged crowns, decay, severe insect or disease damage, or poor vigor. If wildlife enhancement is a goal, you may want to encourage the growth of trees that produce seeds or fruits eaten by wildlife or to thin a dense stand, permitting sunlight to reach the forest floor and stimulate growth of herbaceous plants and shrubs. With aesthetics in mind, you might want to encourage trees with showy spring flowers, brilliant fall color, interesting bark, picturesque shapes, large stems, or other unique features.

As a tree grows it needs more space for its crown to expand and a greater soil volume from which to absorb nutrients and water to sustain fast growth. Thinning redistributes growth potential from many trees to selected crop trees. It also enables crop trees to remain vigorous, thereby resisting insect and disease attacks.

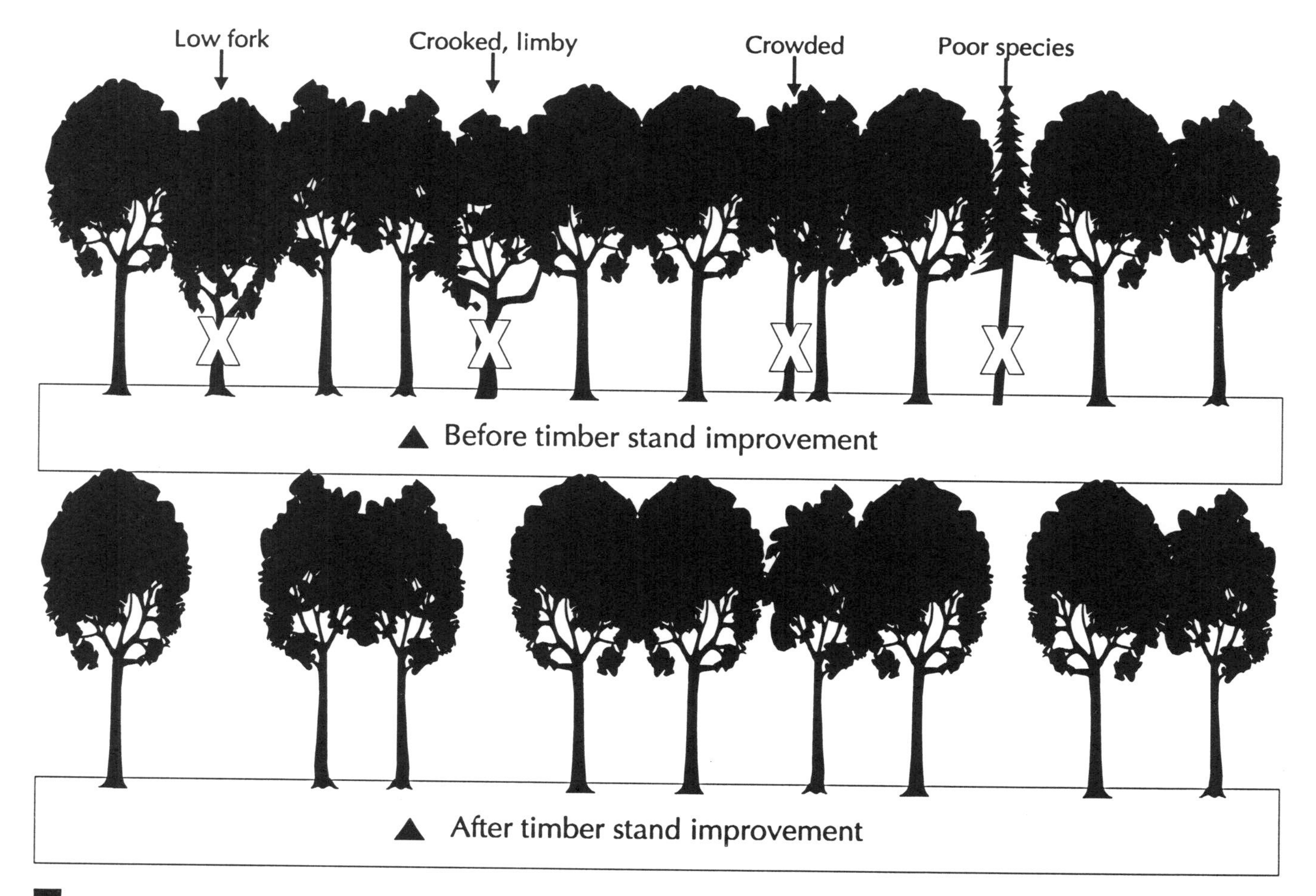

Fig. 31. Timber stand improvement.

The optimum stand density varies somewhat from one forest type to another because of the different growing space requirements of the tree species found in them. A simple method for regulating stand density is to identify crop trees and release them from competition; a more sophisticated method is to apply guidelines from a stocking chart.

When applying the crop-tree release method, identify crop trees that are spaced 20 to 25 feet apart. If crop trees are scarce or unevenly distributed, you can leave two trees as close as 10 feet as long as you treat them as one tree when thinning. Remove trees with crowns that encroach on those of crop trees (Figure 32). Free all sides of sapling and small poletimber-size trees and at least three sides of larger trees. Trees below the main canopy will not affect crop tree growth, but you may cut them if they are marketable. Do not damage crop tree stems and roots while thinning stands. Repeat thinning every fifteen to twenty years.

Stocking charts, which have been developed for many forest types, are useful as thinning guides. Figure 33 is a stocking chart for red pine stands that are managed for maximum fiber production. It assumes there are markets for small-diameter trees. This chart considers basal area, trees per acre, and average stand diameter.

Trees grow best when stocking is between the ***A*** and ***B*** levels. Near the ***A*** level a stand is overstocked; near the ***B*** level a stand is understocked. Generally when a stand approaches the ***A*** level it should be thinned back to near the ***B*** level. For example, if a red pine stand had a basal area of 190 square feet and 350 trees per acre, the average stand diameter would be 10 inches and the stand would be overstocked. By following the straight line for the 10-inch diameter class from the ***A*** curve to the ***B*** curve, we see that trees of this size would grow faster if the stand were thinned back to approximately 85 square feet of basal area and 160 trees per acre. Stocking charts for selected tree species and forest types are shown in Appendix D.

As a rule of thumb, do not remove more than one-third of the basal area from a stand at any one time. A heavier thinning may lead to wind damage and to epicormic branching. Epicormic branches

BEFORE TREATMENT

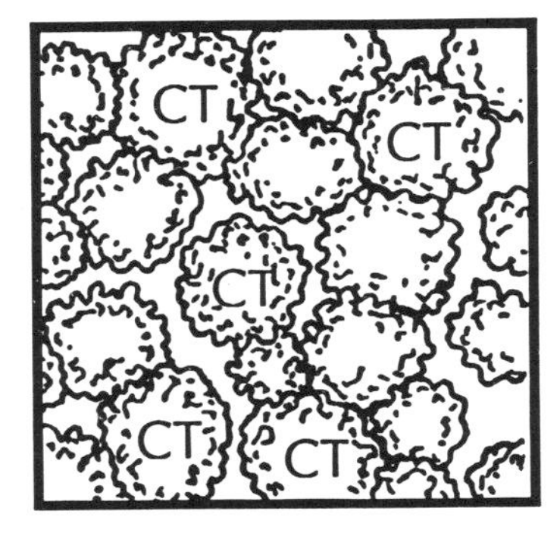

CT = CROP TREE

VIEW FROM ABOVE

VIEW FROM SIDE

AFTER TREATMENT

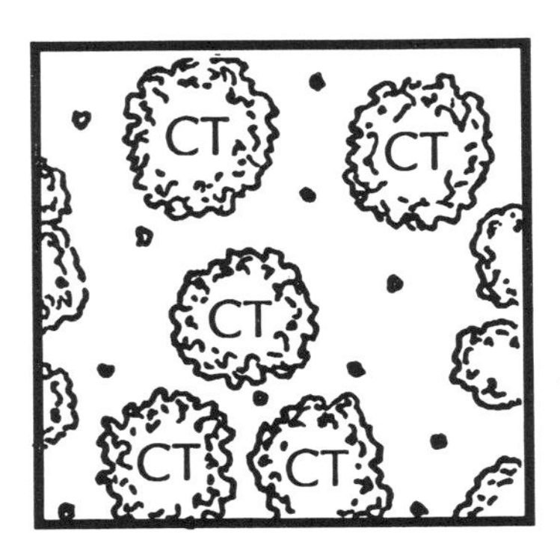

Fig. 32. Crown release will stimulate crop-tree growth. (From *How To Release Crop Trees In Precommercial Hardwood Stands (NE-80-88)*. USDA Northeastern Forest Experiment Station, 5 Radnor Corporate Center, 100 Matsonford Road, Radnor, PA 19087. 2pp.)

(also called water sprouts) arise from dormant buds beneath the bark on hardwood trees. They create knots and lower the quality of wood that can be cut from a tree.

Stand density guidelines for specific forest types are given in Chapter 6 and Appendix D. Methods for determining basal area are explained in Chapter 2.

During a TSI operation, undesirable trees can be killed either chemically or mechanically. If the trees are cut mechanically and re-sprouting is expected to be a problem, treat stumps with a chemical such as 2,4-D (amine), picloram + 2,4-D (amine), or Triclopyr (amine).

When undesirable trees have no commercial value, it often is faster and easier to kill them standing rather than to cut them down. Trees can be killed by girdling, frilling, herbicide injection, or basal spray (Figure 34). Dead standing trees are useful to many wildlife species and they cause little damage to residual trees when they break up. For safety purposes, trees that could fall across roads or trails or on buildings should be felled rather than killed standing.

Girdling involves the complete removal of a 3- to 5-inch band of bark around the trunk with a hatchet or chain saw. Girdling also can be accomplished by encircling the main stem with two parallel chain saw cuts 1 inch deep and 3 to 5 inches apart. This is time-consuming, hard work. The tree may not die for several years and live shoots may arise from below the girdle. Girdling often is used if sprouts are needed for wildlife browse and they will not interfere with more desirable tree seedlings. Girdling is easier in spring when bark separates easily from the wood.

A frill girdle is a single line of downward axe or hatchet cuts that completely encircle the trunk and are filled with 2,4-D (amine), Triclopyr (amine), or picloram + 2,4-D (amine). An oil can with a spout or a squeeze bottle is useful for applying the herbicide. (Do not reuse the container for other purposes after filling it with herbicide.) Treatments during the growing season are more effective than treatments in winter or at the time of heaviest sap flow. Effectiveness also will vary with concentration of the chemical.

An herbicide injection can be used for trees 5 inches in diameter and larger. An injector cuts

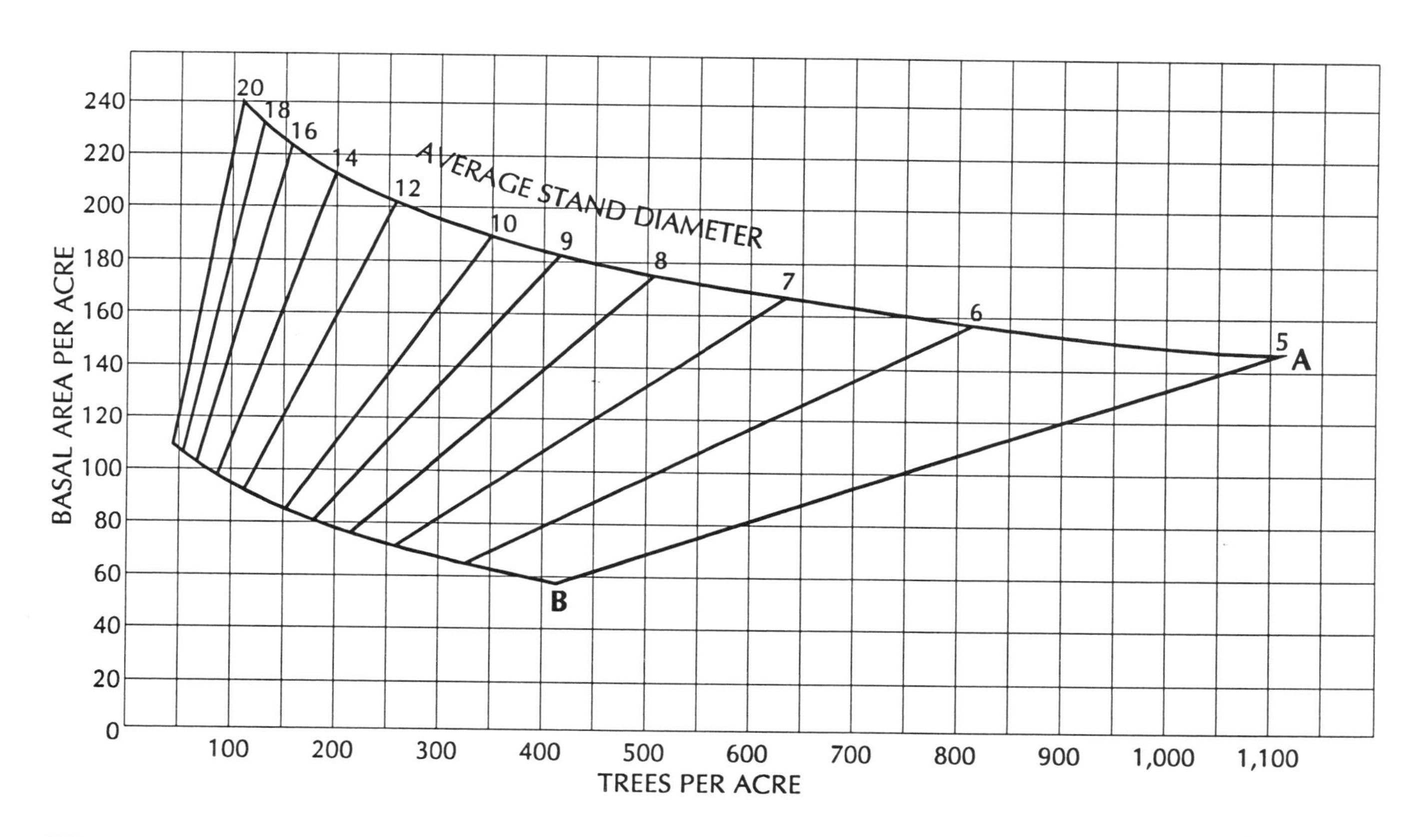

Fig. 33. Stocking chart for red (Norway) pine. (From Benzie, J. W. 1976. ***Manager's Handbook for Red Pine in the North Central States (General Technical Report NC-33).*** USDA Forest Service, North Central Forest Experiment Station, 1992 Folwell Avenue, St. Paul, MN 55108. p. 13.)

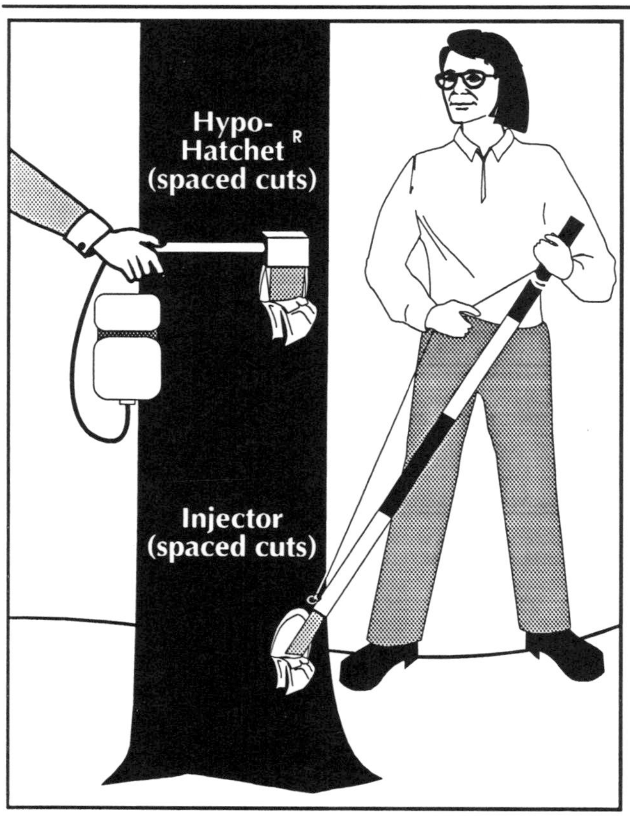

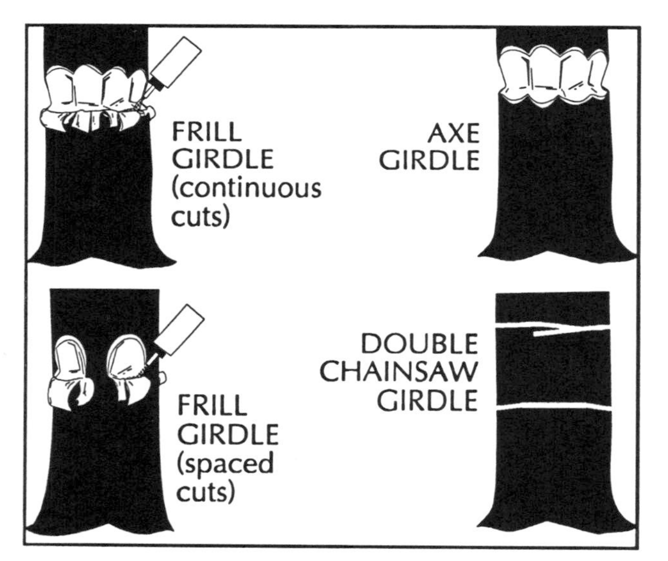

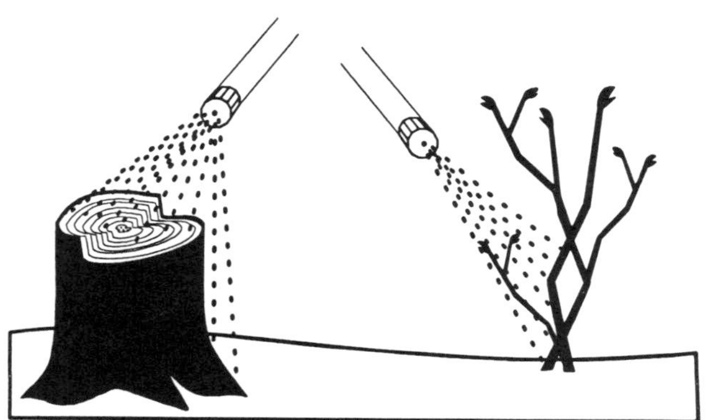

Fig. 34. Methods for killing undesirable trees.

through the bark and injects a measured amount of herbicide into each cut. Cuts usually are spaced 2 to 3 inches apart and are most effective when applied from May though early fall. Herbicides used with injectors include 2,4-D; picloram + 2,4-D; hexazinone, and glyphosate. Some injectors are modified hatchets and others resemble a pipe with a chisel point on one end.

When scattered shrubs or small trees are a problem, you can basal-spray them with a herbicide. Drench the lower 12 to 15 inches of the stem with a herbicide such as Triclopyr or picloram + 2,4-D in oil. Apply chemicals in any season. Control may be poor on root-suckering species. Do not apply chemicals near desirable plants or contaminate surface or ground water.

PRUNING

Individual tree quality can be improved by pruning. Corrective pruning can produce a straight leader, and clear-stem pruning can produce high quality, knot-free wood. Pruning is time consuming and therefore expensive. It should be undertaken with a careful eye on costs. Prune only crop trees—those that have the best form and vigor to produce sawtimber or veneer.

Corrective Pruning

Corrective pruning (Figure 35) will produce a straight stem with one terminal leader. It is done only on high-quality hardwoods, primarily black walnut. Corrective pruning should be done when trees are seedlings or small saplings.

Clear-stem Pruning

Clear-stem pruning (Figure 36) removes lower branches on the stem to produce knot-free wood. Time pruning to coincide with a thinning to maximize knot-free wood production. Pruning is justified when it raises the tree grade enough to increase the stumpage value of the tree beyond the pruning cost. Clear-stem pruning may be appropriate for red pine, white pine, black walnut, red oak, sugar maple, and yellow birch.

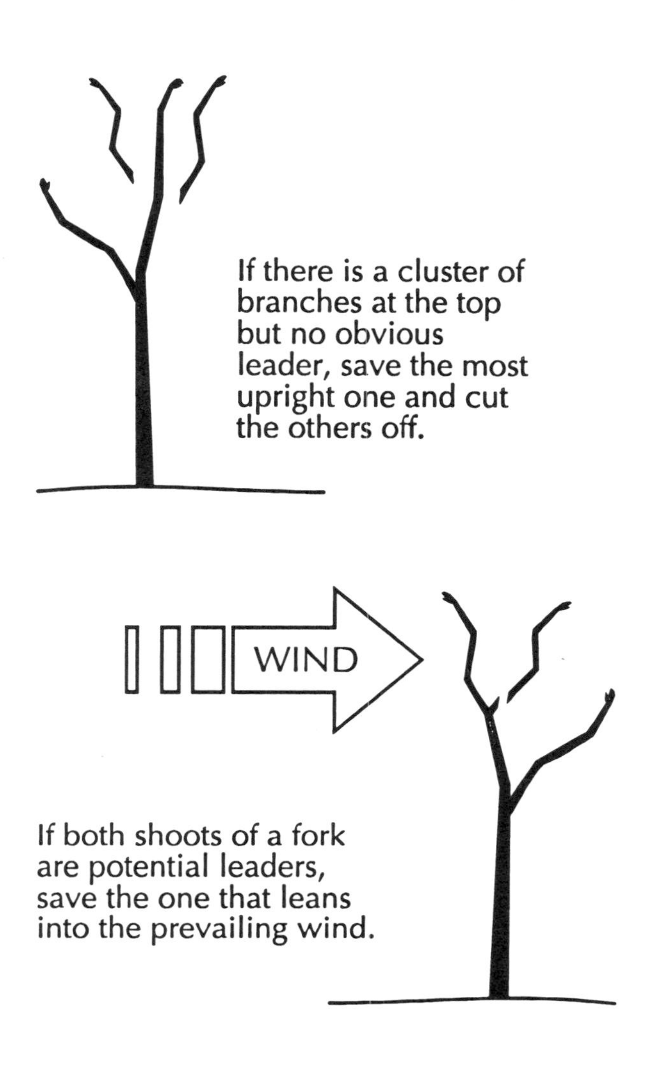

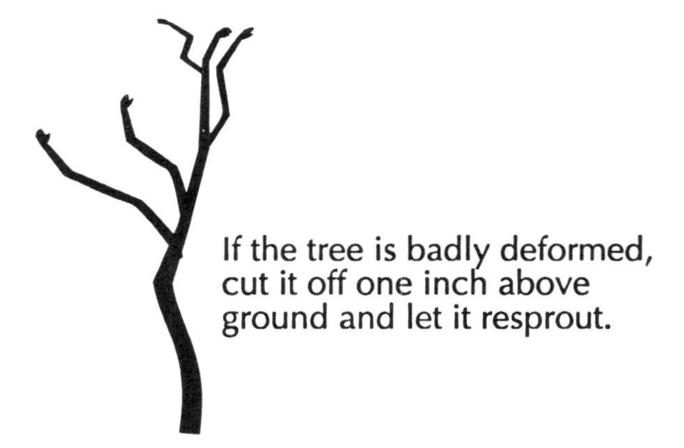

If the tree has a widely forked or multiple top, straighten the most promising leader by using one or more lateral shoots as support:

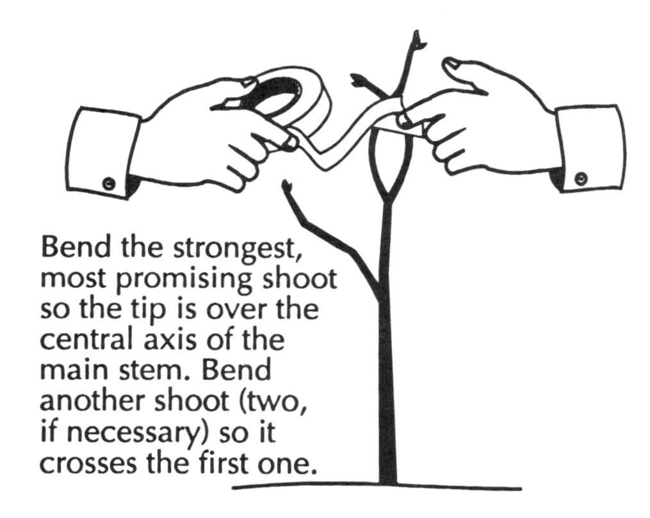

Fasten the shoots together with three wraps of one-inch masking tape. Large or widely divergent shoots may need more layers of tape.

Cut off the tip of the supporting shoot(s) just above the wrapping to eliminate potentially competing new growth.

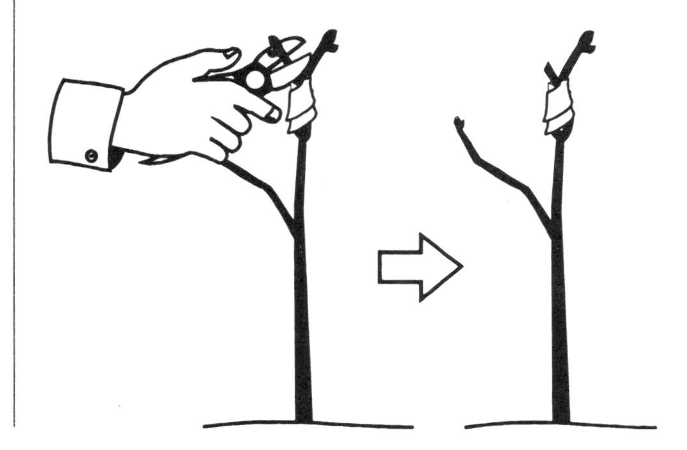

Fig. 35. Corrective pruning.

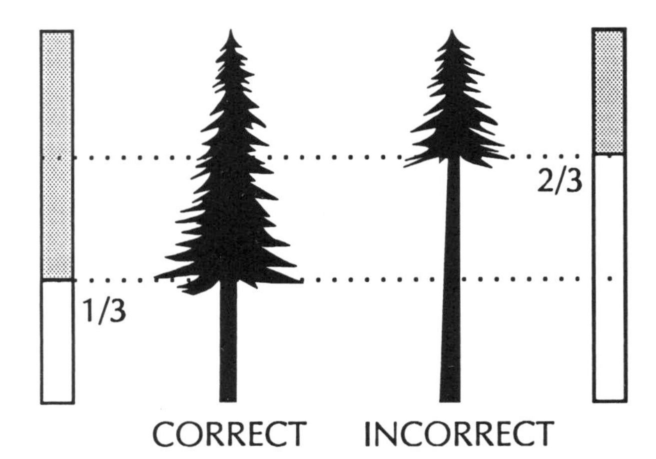

Do not prune higher than 1/3 of total tree height or remove more than 50% of live crown.

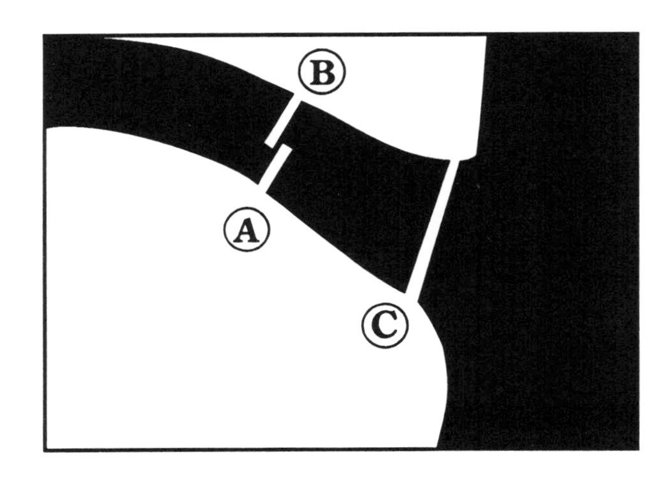

To prune a large limb, cut in order A, B, C.

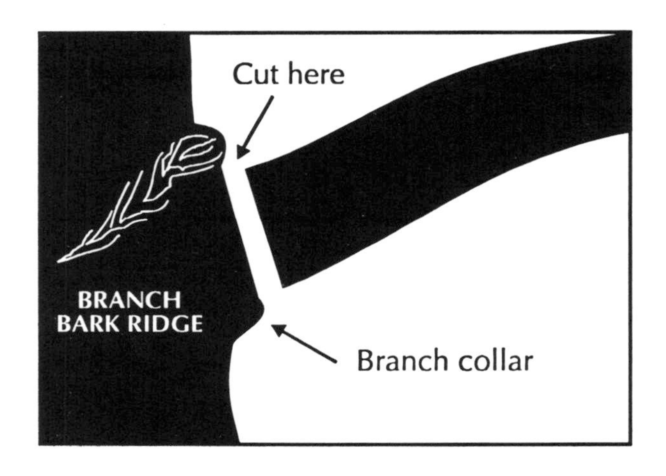

Do not cut into the branch collar or branch bark ridge. Minimize the size of the cut surface.

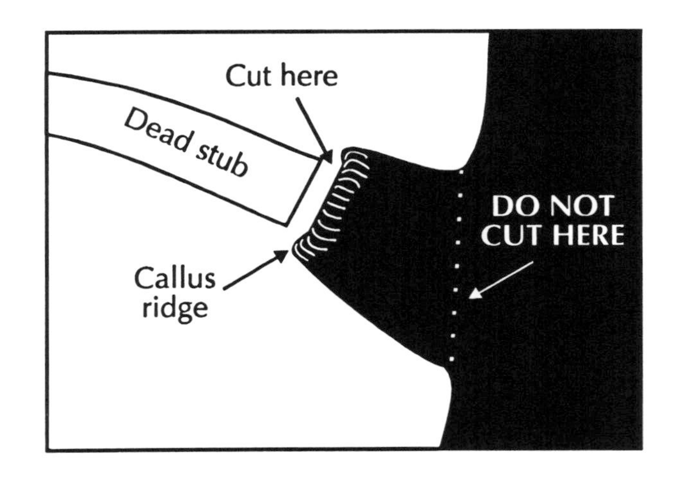

When removing a dead limb, leave callus ridge intact.

Fig. 36. Clear-stem pruning.

Prune trees first when they are 4 to 8 inches average DBH. Prune at least 9 feet high, but no more than the lower one-third of total tree height or 50 percent of the live crown. Additional pruning during the life of a plantation should increase the total clear stem up to 17 feet, if 16-foot sawlogs are the objective.

Prune during the dormant season to help prevent diseases from invading the pruning wounds before they dry out or glaze over with pitch. Do not prune oaks from mid-April through mid-July because oak wilt may enter the pruning cuts. The dormant season also is a more comfortable time to work in the woods because it is cooler, visibility is better, and there are fewer insect pests.

Tree paint and wound dressings do not promote healing. Use them only when an emergency (e.g., storm damage) requires that trees (particularly oaks) be pruned in May or June, or you want to hide the cuts for aesthetic purposes.

When you prune, cut branches close to the trunk, but do not cut into the branch collar—a swelling of the main stem around the base of the branch. Cutting into the branch collar will slow the healing of the pruning cut and may lead to decay. Have a forester show you pruning techniques, especially before pruning young hardwood trees.

Work safely. Wear a hard hat and eye protection. Keep your tools sharp. Sharp tools cut cleanly and require less effort. Tools commonly used for pruning include hand pruning saws, pole saws, shears, and pole pruners (shears). Chain saws are dangerous to use for pruning because you must hold them above shoulder level where they are difficult to control. Chain saws also easily damage trees by cutting into the branch collar or by nicking the bark above or below the pruning cut. Special motorized pruning saws, called tree monkeys, are available in some areas and may be appropriate for large conifer plantations.

Suggested Reference

Lantagne, D. O., R. Heyd, and D. Hall. 1990. *Forest Herbicides for Weed Control in the Great Lakes States (Extension Bulletin E-2219).* Michigan State University, East Lansing, MI 48824. 39 pp.

6

MANAGING IMPORTANT FOREST TYPES AND SPECIES

Tree species that have similar environmental requirements and tolerance for shade and moisture often are mixed together in a woodland. Groups of tree species that commonly occur together are called forest types and are named for the most common tree species in the group.

Each forest type requires somewhat different cultural practices to perpetuate its most desirable tree species and encourage fast growth. This section describes management of important forest types in the Midwest. These descriptions are brief and may not provide adequate information for managing specific stands. We recommend that you have a forester inspect your woodland and prepare stand management plans before you implement any forestry practices.

Range maps and site index curves are included for the predominant species in each forest type. Range maps help you determine what tree species are likely to be found in your area and how well-adapted they may be. Species usually grow better and their wood has greater commercial potential in the heart of the range as opposed to the edge of the range. Site index curves are useful only when you know the site index or average age and total tree height for those species in your woodland. These curves help you determine whether the growth potential on your land is high or low compared to the species' potential.

ASPEN-BIRCH

Products and Uses

Quaking and bigtooth aspen (popple) are grown principally for pulpwood that is used in the manufacture of paper and particleboard, but they also are used for sawtimber and veneer. Paper (white) birch is in less demand than aspen, but can be used for pulpwood, lumber, fuelwood, dowels, and novelty items. Aspen stands are important for ruffed grouse, white-tailed deer, and moose.

Growing Conditions

Both aspen and birch commonly grow in pure stands, but also are found together or in mixtures with other hardwood and conifer species. This section focuses on aspen, which currently is in greater commercial demand than birch.

Aspen will grow on a great variety of soils, but the fastest-growing, best-quality trees often are found on moist soils that are rich in lime and have a high silt-plus-clay content. Trees on dry, sandy soils grow slowly because of low moisture and nutrient levels and are more prone to health problems than trees on better sites. Heavy clay soils do not promote the best growth because of poor soil aeration.

Site index commonly is used for evaluating productivity of different sites when aspen stands are at least 20 years old and have not been damaged by fire or overtopped by other species (Appendix C-1). Only stands with an aspen site index of 60 or better should be used for aspen timber management. Because of slow growth and health problems, poorer sites may be better suited for other species, although some aspen can be grown there for nontimber purposes such as wildlife or aesthetics.

Fig. 37. Range of quaking aspen.

Regeneration

Aspen stands managed for pulpwood generally should be harvested at age 45 to 55. Stands grown primarily for sawtimber generally should be harvested at age 55 to 65. Stands may need to be harvested earlier if disease incidence is greater than 30 percent.

Aspen stands commonly are regenerated from root suckers, but also may regenerate from seed if there is moist, bare mineral soil available during seed dispersal.

Because aspen is very intolerant of shade, optimum root sucker regeneration occurs when the stand is completely clearcut. Do not leave more than twenty mature trees per acre after harvesting the stand. If logging does not destroy undesirable trees and shrubs, remove them by felling, girdling, basal spraying, or controlled burn.

Clearcut when the soil is relatively dry or frozen to avoid damaging the parent roots. This is especially important in stands growing on clay soils with a high water table.

Root suckers grow fastest when trees are harvested during the dormant season and food reserves stored in their roots are at a maximum. Root suckering may not be satisfactory if trees are harvested in the spring or early summer shortly after the leaves and flowers have been produced.

If a stand is harvested during the growing season, root suckers will begin to grow immediately after trees are felled. Do not drive heavy equipment across young sprouts or they will be killed. Avoid damaging new sprouts by beginning the logging activity at the rear of the stand first and progressing toward the log landing.

Old, decadent stands with low vigor and stands with fewer than fifty mature aspen trees per acre may be difficult to regenerate. In these stands you can encourage maximum suckering by harvesting during the dormant season and when the ground is frozen or relatively dry. If an old stand does not have a merchantable volume of wood, you still should fell the trees or shear them using a sharp blade on a bulldozer when the soil is frozen. This will stimulate suckering and help create a better-stocked, new stand.

Two years after clearcutting there should be at least 5,000 aspen root suckers per acre. Some stands may have up to 70,000 root suckers per acre. The more the better; you need many root suckers because aspen stands naturally thin themselves.

If root sucker density following the clearcut appears low, have a forester judge whether the stand is adequately stocked. If it is not, wait at least ten years, then clearcut the stand again. Following this second clearcut, sucker density should improve to a satisfactory level.

Aspen will not compete well in mixtures with other hardwood species such as maple, basswood, ash, and oak. Over time the aspen will die from disease and will be replaced by more shade-tolerant species. Mixed hardwood stands can be managed either for aspen or other hardwoods, but cannot be managed effectively for both. Clearcutting a mixed stand will favor aspen regeneration. Removing aspen during thinnings will favor regeneration of other hardwood species.

In the northern Lake States it is common to find mature aspen with an understory of immature white spruce and balsam fir—two shade-tolerant conifers. These three species can be managed together. If you carefully harvest the aspen and leave the conifers undamaged, openings in the conifer canopy will be large enough to allow good aspen sucker development in scattered patches. In 40 to 50 years the conifers and aspen will mature. Clearcutting then will regenerate a stand of mainly aspen with a few scattered spruce and fir.

Paper birch also is intolerant of shade and usually is regenerated by clearcutting, but recently success has been attained from the shelterwood method. Although small birch produce vigorous stump sprouts when cut, merchantable size trees do not sprout well and sprouts are normally of low quality. Natural seeding from narrow, progressive clearcut strips and small patch clearcuts is the most common regeneration method for birch. It is essential to knock down brush and expose bare mineral soil at the time of clearcutting.

If a clearcut through a birch stand has to be more than 300 feet wide, leave a sufficient number of seed trees scattered throughout the site to get adequate seed dispersal and provide for the survival of seed trees and protection of new seedlings. Remove seed trees within two years after acceptable regeneration.

On hot, dry sites a two-cut shelterwood system may be more successful than clearcutting for birch. The first cut should thin the canopy and provide more sunlight to the forest floor. A year after the first cut, disk the site to lightly bury the birch seeds, help control competing vegetation, and incorporate organic matter. Disking is especially helpful following a good seed fall. After the stand is sufficiently stocked with seedlings, it should be clearcut.

Intermediate Treatments

Once an aspen stand has regenerated, trees grow rapidly. A densely stocked stand thins naturally; artificial thinning is unnecessary to produce pulpwood and may increase losses from hypoxylon canker and rot. Dense stands also promote natural pruning. Artificially thinned stands may produce more sawtimber and veneer than unthinned stands, but thinning should be carried out to grow these products only when disease incidence is low and the site index is 70 or higher. One thinning at about age 30 leaving approximately 240 trees per acre may be appropriate. Take great care to avoid wounding residual aspen trees, since decay and discoloration can enter trees through those wounds.

Pests

Major insect pests of aspen are the forest tent caterpillar, large aspen tortrix, and various borers. Hypoxylon canker and white rot are important diseases.

To reduce losses from these pests, don't grow aspen for timber where the site index is less than 60. Stands growing on poor sites are highly susceptible to pests. Try to regenerate 30,000 suckers per acre and maintain a high number of stems per acre to discourage poplar borer and hypoxylon canker. Do not thin aspen; wounds on residual trees will favor establishment of poplar borer and hypoxylon canker.

To minimize pest damage, harvest trees by age 40 unless the site index is at least 75 and veneer is the desired product. If fewer than 15 percent of the trees are infected with hypoxylon canker, stands can be permitted to grow longer than 40 years. If 15 to 25 percent of the trees are infected with hypoxylon canker, harvest early and regenerate aspen. If more than 25 percent of the stand is infected, consider converting to an alternate forest type or species. Harvest early if white rot affects more than 30 percent of the basal area.

Repeated defoliation by forest tent caterpillar will weaken the trees, increasing their susceptibility to disease. Insecticides may be required to protect the stand during prolonged outbreaks.

BLACK SPRUCE

Products and Uses

Black spruce is grown almost exclusively for pulpwood. In the past, Christmas trees were cut from the top 3 feet of old trees that were 20 to 35 feet tall and growing on poor sites. The spruce grouse depends on the black spruce type for most of its habitat needs.

Growing Conditions

Black spruce grows mainly in pure stands, but also may be mixed with tamarack, northern white-cedar, and balsam fir on organic soil and with other conifers, especially jack pine, and various hardwoods, including black ash, red maple, and paper birch, on better drained, mineral soil.

The black spruce forest type is found mainly on organic soil in the Lake States, but black spruce trees also commonly occur on upland mineral soil in mixture with other species. The best black spruce sites on organic soils occur where the soil water is part of the regional groundwater system and is enriched by nutrients flowing from mineral soil areas. They have moderately well-decomposed organic soil that contains much partially decayed wood and is dark brown to black. However, the upper 6 inches may be poorly decomposed sphagnum or other mosses, especially in old stands. The poorest sites occur where the soil water is separated from the groundwater system, and where there is 2 feet or more of poorly decomposed, yellowish-brown sphagnum moss.

Black spruce is common on mineral soil only on the Laurentian Shield in northeastern Minnesota and in a few areas of Upper Michigan. Growth is best where the slope is gentle and moisture is plentiful either from a shallow water table or seepage along bedrock.

Site index curves for black spruce are shown in Appendix C-2.

Fig. 38. Range of black spruce.

Regeneration

Rotation lengths for black spruce range from 60 to 140 years, but where dwarf-mistletoe is a problem, rotations usually should not exceed 100 years on organic soils or 70 years on mineral soils.

Black spruce stands 40 years or older have a nearly continuous seed supply because cones remain on the trees and shed their seed over several years. In addition, seed crops seldom fail. Reproduction by layering is common, particularly in swamps and bogs.

Clearcutting blocks or strips is the best method for harvesting and reproducing black spruce. Seedling establishment requires a seedbed that is moist but not saturated and free from competing vegetation. Establishment is generally successful if the surface layer of the soil is:

1. Removed by fire or machine;
2. Compacted, as in a skid road; or
3. Composed of living sphagnum moss.

Feather mosses make poor seedbeds because they dry up and die after clearcutting. Broadcast burn heavy slash if it covers desirable residual trees or good sphagnum moss seedbeds. Skidding whole trees from the site may eliminate the need for slash disposal. If your site requires broadcast burning, you can reduce the cost per acre by clear-cutting and direct seeding large blocks (40 acres or more).

Sow seed between March and mid-May of the first year following burning. On well-prepared sites, 2 to 3 ounces of seed per acre should be adequate. Seed should be treated with bird repellent and fungicide.

Natural seeding can be effective with large, wind-firm stands. These sites may not require broadcast burning. Cut progressive strips perpendicular to the prevailing wind to maximize seed dispersal and minimize wind damage. Successive strips should be cut into the wind. A strip perpendicular to the prevailing wind can be up to 6 chains wide (1 chain = 66 feet) if natural seeding can occur from both sides or 4 chains wide if natural seeding can occur only from the windward side. The outer portion of large stands can be reproduced by natural seeding, thus significantly reducing the area requiring direct seeding.

Natural seeding of black spruce, especially on nonbrushy sites, often results in stands that are too dense for optimum pulpwood growth. To avoid overstocking, count the trees about three years after site preparation. If there are at least six hundred healthy, well-spaced black spruce seedlings per acre that are at least 6 inches tall, clearcut the adjacent area of mature spruce to eliminate further seeding into the new stand.

Planting seedlings is more reliable than seeding, but also more expensive. Black spruce can be planted successfully using 3-0 or container-grown seedlings. Transplants are expensive but useful where serious weed competition is expected.

Intermediate Treatments

Thinning of overstocked sapling and poletimber stands is generally not economical and may lead to increased wind damage. Although black spruce is shade tolerant, on good sites a dense overstory of undesirable shrubs or hardwoods may severely suppress seedling growth. In these situations, control brush to release the spruce.

Pests

Eastern dwarf-mistletoe is the most serious disease affecting black spruce. It causes branch deformations (witches' brooms), reduces growth, and eventually kills trees. Since mistletoe survives only on living trees and spreads slowly, you can control it by killing all trees in infected areas plus a border strip 1 to 2 chains wide, then burning the site with a hot fire. To avoid mistletoe infections, clearcut and burn all mature stands where feasible to eliminate undetected mistletoe sources.

Wind may cause substantial losses in older black spruce stands by breaking or uprooting trees, especially where butt rot is present and where stands have been opened up by partial cutting. Minimize wind damage by shortening rotations and clearcutting narrow strips that progress over time toward prevailing winds.

Spruce budworm is not a common problem on black spruce because of the spruce's late budbreak.

BLACK WALNUT

Products and Uses

Wood products from black walnut include sawlogs, veneer logs, gun stocks, and smaller novelty pieces. Frequent nut crops make it an excellent tree to plant for wildlife, especially squirrels. In some areas nuts are collected and sold for human consumption.

Growing Conditions

Black walnut generally is found scattered among other tree species. Pure stands are not common, but do occur. Walnut grows best on lower north-and east-facing slopes, stream terraces, and floodplains. It is common on limestone soils and grows well on deep loams, loess soils, and alluvial deposits that are fertile and moist but well drained. Poor sites for walnut include steep south- and west-facing slopes, narrow ridgetops, and poorly drained sites. Soils with acid clayey subsoils, coarse sand or gravel layers, or bedrock within 2-1/2 feet of the surface are not suitable for walnut. Black walnut leaves and roots actively secrete material toxic to some trees, shrubs, and herbaceous plants.

Site index curves for black walnut plantations are shown in Appendix C-3.

Regeneration

The rotation length for black walnut is from 50 to 80 years. Black walnut can be regenerated naturally from seed and stumps sprout well if trees are less than 20 to 30 years old. Since black walnut trees normally are a minor component of a woodland, natural regeneration is seldom reliable. Seedlings should be planted to supplement natural regeneration.

Seedlings often are planted at a spacing of 10 x 10 feet for timber production purposes, and 15 x 15 feet for producing a combination of timber and nuts. Black walnut will not tolerate shade. To prepare a woodland site, cut and/or kill with herbicides all woody vegetation larger than 1/2-inch in diameter. Grassy and weedy sites may require herbicide treatments to kill existing vegetation prior to planting seedlings.

Plant seedlings in the spring as soon after the ground thaws as possible. Seedlings at least 1/4-inch in diameter at 1 inch above the root collar will compete better with weeds and have a better survival rate than smaller seedlings.

Seeds are easier and less expensive to plant than seedlings, but should be protected from squirrels and other rodents. Mechanical barriers (hardware cloth, tin cans, etc.) are most reliable, but they are expensive and time consuming to install.

Seed can be sown in either the fall or spring. You do not need to remove the husks for fall planting. Spring planting eliminates overwinter feeding by rodents, but requires that the seed be stratified before planting to break dormancy. Stratification involves subjecting seed to cold temperatures and regulating moisture for a period of time.

Intermediate Treatments

Control weeds for at least three years when establishing a walnut plantation to maximize the sunlight, moisture, and minerals available to walnut seedlings and to reduce plant cover that encourages rodents. You can control weeds by mowing or cultivation in open field plantings, but in most situations herbicides are more cost effective and will not damage the walnut stems or roots.

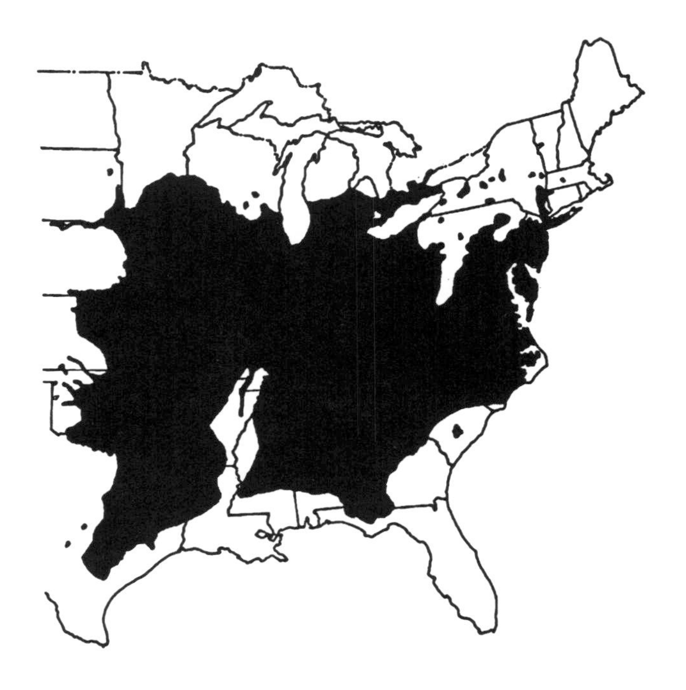

Fig. 39. Range of black walnut.

Corrective pruning can improve seedling form if tip dieback or stem forking has occurred. Do not prune too heavily; young stems have a strong natural tendency to grow upright. Clear-stem pruning is recommended to help produce knot-free wood (see Chapter 5).

Fertilization generally is not needed on a good black walnut site unless a specific nutrient element is deficient. Foliage analysis will reveal a nutrient deficiency. Weeds are the usual benefactors of fertilizers.

Thin the stand lightly and frequently—perhaps every 10 years—to maintain rapid, uniform growth. Following thinning at least three-fourths of the crown of crop trees should be 5 feet or more from the crowns of adjacent trees. Competing trees should be cut or girdled and treated with a herbicide to prevent resprouting. Dominant and codominant trees will respond best to thinning.

Pests

The major pests of black walnut are walnut caterpillars, bud borers, anthracnose, and fusarium canker. Pesticides usually are not economical. Fungicides may be necessary to control anthracnose for the purpose of improving nut production, and insecticides may be necessary to control caterpillars. Anthracnose can be managed by controlling weeds that weaken the trees. Fusarium canker can be controlled by restricting pruning to the late winter. Fire is highly damaging to black walnut.

Incorrect pruning can lead to serious problems, including fusarium canker, bark necrosis, and sunscald. Refer to Chapter 5 to learn more about proper pruning procedures.

BOTTOMLAND HARDWOODS

Products and Uses

Elm, ash, cottonwood, silver maple, and associated hardwoods are used primarily for lumber, veneer, and firewood, but in some areas cottonwood also is used for pulpwood. A high number of wildlife species, especially birds, can be found in a mature bottomland hardwood forest. Mature and overmature stands provide cavities essential to many wildlife species including woodpeckers, wood ducks, barred owls, and raccoons. White-tailed deer and beaver also can be found in this forest type.

Growing Conditions

Species composition of bottomland hardwoods varies depending on the site, but may include American elm, green ash, black ash, eastern cottonwood, silver maple, and black willow. Ash, cottonwood, and silver maple are the most commercially desirable species.

Ash is common on alluvial soils along rivers and streams. It is found where good moisture conditions occur in medium- to coarse-textured sands or loams to clays. Black ash can grow on wetter sites than green ash. Ash seedlings become established in partial shade on moist litter or mineral soil.

Fig. 40. Range of green ash.

Eastern cottonwood is common on moist alluvial soil ranging from coarse sands to clays but grows best on fine sandy loam near streams. Cottonwood seedlings require moist, exposed mineral soil and full sunlight for establishment.

Silver maple is found on alluvial flood plains of major rivers where there are moist, fine-textured silt and clay soils that are imperfectly drained. Silver maple seedlings require moist litter or mineral soil and full sunlight for establishment.

Stands with a site index below 70 for green ash (see Appendix C-4), eastern cottonwood (see Appendix C-5), or silver maple should be managed for wildlife, aesthetics, or other nontimber uses.

Regeneration

Bottomland hardwood species can regenerate from their light, wind-disseminated seeds. Some species, especially ash, also may stump sprout. Clearcut all trees greater than 2 inches DBH to promote regeneration. Tree seedlings pre-

Fig. 41. Range of eastern cottonwood.

DBH to promote regeneration. Tree seedlings present in the understory prior to harvest (called advance reproduction) usually are not abundant, and if present, will probably consist of elm, maple, and possibly ash. A dormant season harvest encourages more stump sprouts, but should be planned to scarify the soil surface, providing exposed mineral soil for seed germination.

Consider planting within two years after harvest if natural regeneration is not adequate. Since planting is expensive, confine planting to the best sites and avoid locations that frequently flood. Prepare the site as follows:

1. Shear all residual woody vegetation near ground level.
2. Pile debris in windrows and burn it.
3. Rake the entire surface to collect any remaining vegetation.
4. Deeply disk and till the planting bed.

Cottonwood generally is established from cuttings. Ash generally is established from seedlings. Plant at a 12 x 12-foot spacing as early as possible in the spring using genetically improved stock. An auger will give better survival than a planting bar. Mechanical weeding will be necessary for the first two or three years.

Intermediate Treatments

In very dense stands a precommercial thinning and weeding of undesirable trees is recommended to concentrate growth on the most desirable trees. The first commercial thinning should occur when codominant trees average 8 to 10 inches DBH. Two or three more thinnings will be required every 7 to 15 years to sustain fast growth. Remove diseased trees and those of low vigor or poor form. Follow the crop-tree release method described in Chapter 5 or the stocking chart for elm-ash-cottonwood in Appendix D-1 to determine when and how much to thin. At final harvest most stands should have 120 to 130 square feet of basal area (roughly 50 high-quality trees) per acre of commercial species.

Pests

Major insect pests in bottomland hardwoods are the forest tent caterpillar and cankerworms. Chemical or microbial insecticides may be required to control these defoliators. Major diseases include Dutch elm disease, ash yellows, and cytospora canker. Harvest commercial-size elms whenever possible to salvage their value before Dutch elm disease kills them. Retain elms during thinning only when no other desirable tree is available. Reduce canker damage by thinning to promote tree vigor, but be very careful to avoid damaging residual trees.

EASTERN WHITE PINE

Products and Uses

Eastern white pine is used mainly for lumber. Although wildlife receive virtually no food benefit from older white pine stands, young dense stands may be used for shelter during inclement weather. Deer and rabbits browse heavily on white pine seedlings. In contrast to red pine stands, white pine (except in pure, dense stands) tends to have a better developed tall shrub and herb layer providing wildlife habitat.

Growing Conditions

While white pine occurs in pure natural stands, it more frequently is a component of other forest types that may include red pine, red oak, hemlock, ash, and other northern hardwoods or conifers. It is most common on well-drained sandy soils and on droughty, loamy sands where it competes well with hardwoods. White pine grows well on stony loams, silty loams, and glacial tills with good or impeded drainage, but usually cannot compete with more aggressive hardwoods. The worst soils for white pine are clayey or poorly drained. Aspect and slope seldom restrict its occurrence.

Site index curves for eastern white pine in the Lake States are shown in Appendix C-6.

White pine is intermediate in shade tolerance and may live 200 years or more. Seed production is generally good every three to five years, although 10 years or more may elapse without a good seed crop if cone beetle damage is present.

Fig. 42. Range of eastern white pine.

Regeneration

Rotations may be as long as 120 years for white pine, but 80 years usually is sufficient to produce sawlogs on good sites. Seedlings grow best under partial shade. Seedlings require at least 20 percent of full sunlight to survive, but may die from high soil surface temperature if there is full sunlight. White pine reproduces naturally from seed. Seeds will disperse at least 200 feet within a pine stand and more than 700 feet in the open.

A two-cut shelterwood system probably is the most reliable method for natural regeneration. Ten years prior to the final harvest remove 40 to 60 percent of the overstory (no more than 30 to 40 percent of the basal area), preferably in the year before or during a good seed year. Harvest during snowless months to scarify the site and expose mineral soil. Remove hardwood regeneration during the harvest since these trees may seriously compete with pine seedlings. After 5 to 10 years, if white pine seedlings are abundant, clearcut the residual overstory. (Delay this harvest until the new white pines are 20 to 25 feet tall if white pine weevil is expected to be a problem.) If white pine regeneration is not satisfactory, you may need to again thin the overstory, control advance hardwood regeneration, and wait another 5 to 10 years before the final harvest. Consider planting white pine seedlings to increase the density to 500 to 600 seedlings per acre.

Mechanical site preparation and planting are required on bare land or in white pine stands that do not naturally regenerate. Plant 2-0 or 3-0 seedlings at rates up to 600 to 800 trees per acre (closer where heavy white pine weevil damage is expected). Plant under a light forest canopy to reduce weevil and white pine blister rust damage.

Intermediate Treatments

You do not need to thin white pine seedling and sapling stands, but if a hardwood overstory develops, partially remove it to maintain 50 percent of full sunlight on the white pine. When trees average 6 to 8 inches DBH, begin thinning and remove the hardwood overstory. Use the stocking chart for eastern white pine in Appendix D-2 as a thinning guide. When stands reach the **A** level, cut them back to the **B** level. Basal area after thinning should be about 100 square feet for young stands and 150 square feet for older stands. Maintain at least a 35 percent live-crown ratio on crop trees. Since white pine has persistent branches, you will need to prune to a height of 17 feet to develop clear wood. Prune in the dormant season, removing limbs only if they are less than 2 inches in diameter. Frequent light prunings are preferred to a single heavy pruning. Depending on local markets, pruning may not be economical.

Pests

White pine blister rust is a serious disease problem. Major insect pests include white pine weevils, bark beetles, and pine sawflies.

Your local forester can advise you concerning the blister rust hazard in your area. Don't plant white pine in high-hazard zones for blister rust. In medium and low-hazard zones, prune lower branches early to minimize the disease. Weevil and blister rust damage can be reduced by regenerating white pines under an overstory of hardwoods and releasing them slowly until the pines are about 20 to 25 feet tall. Then remove the overstory.

To avoid bark beetle damage, do not harvest or thin the stand from January through August unless you destroy all slash greater than 2 inches diameter within three weeks of cutting. Slash can be destroyed by piling and burning, by chipping, or by burying. Avoid wounding residual trees during thinning. If trees are damaged by fire, storms, or logging, harvest them, remove the logs from the woodland, and destroy the slash within three weeks.

Sawflies are common defoliators on white pine, but usually do not cause serious damage. If control is needed, chemical sprays will be required, since microbial insecticides are ineffective.

JACK PINE

Products and Uses

Jack pine is used mainly for pulpwood, but also for poles and small sawlogs. It is moderately useful for deer food. Young trees may be heavily browsed where deer populations are high. Dense, young stands provide cover for snowshoe hares. Dense sapling and poletimber stands offer some wildlife shelter, but not as much as most other conifers. Older jack pine stands usually are less dense than other conifer stands, permitting the growth of understory shrubs and herbaceous plants offering wildlife food and cover.

Growing Conditions

Jack pine grows in extensive pure stands, but frequently is mixed with red and white pine, aspen, paper birch, and oak; less often it is mixed with black spruce, white spruce, and balsam fir. It grows best on well-drained loamy sands where the midsummer water table is 4 to 6 feet deep, but it is more common on dry sandy soils where it grows poorly, but better than most other species. It also is found on glacial eskers, sand dunes, and rocky soils. Jack pine does well on moderately acidic soils, but it will tolerate slightly alkaline conditions. It grows poorly in areas of shallow bedrock and heavy clay soil. A shade-tolerant species, jack pine is a pioneer species that typically colonizes burned areas and bare mineral soil. It usually is succeeded by more shade-tolerant species on all but dry, sandy soil.

Site index curves for jack pine are shown in Appendix C-7.

Regeneration

Although jack pine is short-lived, stands sometimes survive up to 100 years. Individual trees may live for 200 years on good sites. Commercial rotations generally are 40 to 70 years. Mature trees in these stands range from 8 to 12 inches DBH.

On all but dry sandy soils, species other than jack pine are more productive and valuable for wood products. If your management goal is timber production, stand conversion to alternate species is recommended on these sites. In mixed species stands sufficiently stocked with more desirable species, you can make the conversion by harvesting the jack pine in several cuts. If desirable species are not well-stocked, convert the stand by clearcutting and planting alternate species.

Jack pine cones in parts of the range are **serotinous,** that is, they remain closed at maturity. They open only when exposed to intense heat (such as heat from a fire), but persist on the trees for years resulting in large accumulations of seed.

Fig. 43. Range of jack pine.

In the southern part of the range, cones typically open as soon as they mature. Good seed crops occur at 3- to 4-year intervals starting about age 20, but the best seed production occurs on trees 40 to 50 years old.

Jack pine seedlings require full sunlight. Clearcutting creates the best conditions for regeneration, but seed tree or shelterwood systems may be appropriate depending on stand and site conditions. Clearcutting is recommended where a new stand will be established by planting improved seedlings, direct seeding, or scattering serotinous cones from high-quality trees. After clearcutting, branches bearing serotinous cones can be scattered on bare mineral soil. The sun's heat near the ground surface will open the cones and release the seed. If the mature stand is not a suitable seed source, burn the site to destroy slash and plant or seed the area using a desirable seed source.

The seed-tree system is a possible alternative for stands that have ten well-distributed, desirable quality seed trees per acre with an abundant supply of nonserotinous cones. After the harvest, burn the area to consume slash, kill competition, and prepare a favorable seedbed. Burn slash as soon as possible after harvest to minimize the risk of seed trees windthrowing before they cast seed. Jack pine slash requires a month of warm, dry weather to cure sufficiently to burn. Early spring fires permit seeding during the most favorable season, but late fall burning and seeding may be almost as effective if rodent populations are low.

Direct seeding may be successful where the water table is within a few feet of the surface or there is frequent precipitation during germination and early seedling development. Coat the seed with bird and rodent repellents and sow it at the rate of 20,000 viable seeds per acre. It's best to seed in early spring to take advantage of snowmelt waters and spring rains.

You may need to plant where direct seeding failed or on deep, dry sandy soils. Bare-root seedlings should be planted only in spring, but container-grown stock can be planted into early summer. A 6- to 8-foot spacing usually is recommended.

Intermediate Treatments

Most natural jack pine stands are understocked; however, dense seedling or sapling stands may develop that will stagnate if not thinned. In very dense seedling stands (e.g., 10,000 trees per acre) it's less expensive to mechanically clear strips about 8 feet wide and leave strips about 2 feet wide than to thin to produce 800 to 1,000 uniformly spaced crop trees per acre. Thin pole-sized stands according to the stocking chart for jack pine in Appendix D-3 if poles or sawtimber are desired. Because pulpwood is the main crop, pruning is not recommended.

Pests

Common insect pests of jack pine include bark beetles and jack pine budworm. Heart rot and shoestring root rot are important diseases.

To reduce losses, harvest stands by 50 years of age. (If the site index is greater than 70, you can grow the trees longer to produce sawlogs.) Do not reproduce jack pine where the site index is less than 55. To maintain vigorous stands on good sites, thin regularly, removing suppressed and low-vigor trees while avoiding damage to residual trees. This reduces flower production and helps control both budworms and bark beetles.

To minimize budworm buildup, avoid open stands, open-grown wolf trees, and stands with suppressed trees. If budworms kill the tree tops, harvest the stand within a few years to avoid loss due to heart rot. Minimize bark beetle damage by maintaining vigorous stands, avoiding tree damage, and managing slash as described for eastern white pine. There is no treatment for shoestring root rot.

NORTHERN HARDWOODS

Products and Uses

The northern hardwood forest type includes numerous tree species. Sawlogs and veneer logs are the major wood products, but some species also are harvested for pulpwood. Maple syrup is made from sugar maple sap. Wildlife found in a northern hardwood forest may include deer, bear, squirrel, ruffed grouse, and woodcock.

Growing Conditions

Species composition of a northern hardwood forest varies by site and geographic range. Species may include sugar maple, American basswood, white ash, black ash, yellow birch, red maple, and elms. Occasionally aspen, paper birch, balsam fir, and northern red oak are important. Beech and eastern hemlock occur from Michigan eastward.

Sugar maple, beech, hemlock, and balsam fir are very shade tolerant; basswood is tolerant; yellow birch, white ash, red maple, and red oak are moderately tolerant; black ash, paper birch, and aspen are intolerant. Elms, black ash, yellow birch, red maple, eastern hemlock, and balsam fir survive best on high-moisture sites. Sugar maple, white ash, basswood, and beech generally are confined to better drained soils. The best timber is found on moist, well-drained, fertile, loamy soil. The poorest sites occur on soils that are infertile, dry, shallow, or swampy.

Site index comparisons among hardwoods are shown in Appendix C-8.

Fig. 44. Range of sugar maple.

Regeneration

Northern hardwoods can be regenerated by a wide range of systems depending on the species to be favored. If high-quality, very shade-tolerant species are desired, use single-tree selection or group selection methods. After a selection harvest, the residual basal area should be approximately 80 square feet per acre distributed among tree size classes as shown in Table 8.

If you prefer an even-aged stand dominated by sugar maple, use a two-cut shelterwood system. Harvest in winter and leave 60 percent crown cover after the first harvest. Make the second cut after advance regeneration is 2 to 4 feet high. If you prefer a greater variety of species, use a two-cut shelterwood system, but first eliminate all reproduction present before cutting, harvest in any season except summer, scarify the site during harvest, leave 70 to 80 percent crown cover, remove undesirable seed sources, and make the second cut after advance regeneration is 2 to 4 feet high.

Planting seedlings is rarely necessary, but is appropriate for open fields or under a shelterwood to change the species composition. In open fields plant only in fertile, well-drained soils. Thoroughly disk before planting, plant tap-rooted species such as white ash and northern red oak, plant only when there is good soil moisture, and control weeds for 1 to 3 years after planting. Under shelterwoods, kill undesirable understory plants and plant in the most open areas immediately after site preparation.

Where aspen is mixed with more shade-tolerant northern hardwood species, decide whether to encourage either aspen or the other species. If there is an overstory of aspen and an understory of hardwoods, you can favor the aspen by clearcutting the stand to stimulate root suckering. Favor hardwoods by removing the aspen when the understory hardwoods are 1 to 3 inches DBH, taking great care to avoid damaging the hardwoods. If the aspen has little commercial value, consider killing it with herbicides and letting it stand.

If aspen and other hardwoods are of equal size, aspen can be favored by clearcutting the stand. If aspen is scarce but desirable, follow the harvest with burning or shallow scarification to create an aspen seedbed. To encourage hardwoods, thin or harvest the stand according to the stocking chart for even-aged management of northern hardwoods (Appendix D-4).

Intermediate Treatments

When following the single-tree selection system, use Table 8 to determine the approximate basal area and number of trees to leave after each harvest. Remove poor quality trees and undesirable species during the harvest. When even-aged stands are created through shelterwood harvests or clearcuts, follow the stocking chart in Appendix D-4.

Pests

Insect pests vary due to the diverse species composition of the northern hardwoods forest type. Forest tent caterpillar and a fall defoliator complex (a mixture of up to ten insect species) cause the most problems. No cultural controls are available. Chemical insecticides and BT are effective.

Nectria canker can be common, especially in uneven-aged stands. Reduce damage by maintaining healthy stands and removing infected stems. Sapstreak of maple and heart rots also are serious. To minimize damage from these diseases, reduce damage to roots and stems during cutting operations by using rubber-tired skidders and by harvesting during winter or dry seasons. Remove diseased trees as soon as possible.

Table 8. Desirable stocking per acre for uneven-aged management of northern hardwoods.

RESIDUAL		
DBH IN INCHES	NUMBER OF TREES	BASAL AREA IN SQUARE FEET
5	21	2.9
6	15	2.9
7	12	3.2
8	9	3.1
9	8	3.5
SUBTOTAL	65	16.0
10	7	3.8
11	6	4.0
12	5	3.9
13	5	4.6
14	5	5.3
SUBTOTAL	28	22.0
15	4	4.9
16	4	5.6
17	3	4.7
18	3	5.3
19	3	5.9
SUBTOTAL	17	26.0
20	2	4.4
21	2	4.8
22	2	5.3
23	1	2.9
24	1	3.1
SUBTOTAL	8	20.0
TOTAL	118	84.0

SOURCE: Hutchinson, J. A. (Ed.). *Northern Hardwood Notes (Note 4.03).* U.S. Government Printing Office, Washington, DC 20402.

NORTHERN WHITE-CEDAR

Products and Uses

The demand for good northern white-cedar lumber is strong, but many mature stands do not have enough volume for a commercial harvest. White-cedar also is used for fence posts and poles. The white-cedar forest type is valuable for deer yards, but some have inadequate shelter or browse. Deer habitat is best if white-cedar stands at different stages of development are interspersed throughout the forest.

Growing Conditions

Northern white-cedar grows in pure stands but is more common in mixed stands. Common associates on wetter soils include balsam fir, black spruce, white spruce, tamarack, black ash, and red maple. On better drained and upland soils, white-cedar may be found with aspen, eastern white pine, eastern hemlock, yellow birch, or white birch. It may perpetuate itself in pure stands, but other tree species seem to gradually replace it in mixed stands, particularly after disturbances. It grows best on limestone-derived soils that are neutral or slightly alkaline and moist but well drained. It also grows well on well-decomposed, neutral or slightly alkaline organic soil derived from woody plants or sedges and on organic soils in which the upper 4 inches are poorly decomposed sphagnum or other mosses. The best sites have moving soil water and usually are near streams or other drainageways. The poorest sites have poorly decomposed acid soil throughout the whole root zone that is derived from plants such as sphagnum moss. These sites have little water movement (except during snowmelt) and often are far from drainageways.

Site index curves for northern white-cedar are shown in Appendix C-9.

Rotation lengths range from 70 years for posts up to 160 years for poles or small sawlogs. When stands are managed for deer shelter, rotations should run at least 110 years.

Fig. 45. Range of northern white-cedar.

Regeneration

Northern white-cedar is shade-tolerant and can be managed under the single-tree selection or clearcutting systems. A clearcut or shelterwood harvest followed by natural seeding is the usual regeneration method. If advance reproduction is not present, a combination of clearcut and shelterwood strips is recommended to optimize natural seeding. Strips vary from 1 chain wide where seed-bearing trees are less than 35 feet tall to 2 chains where these trees are more than 60 feet tall. Use either alternate or progressive strips. If you use alternate strips, clearcut one set, then cut the adjoining strip in two stages using the shelterwood system about 10 years later. For the first stage of the shelterwood, leave a basal area of 60 square feet per acre in uniformly spaced dominant and codominant trees of desirable species. Select residual trees for good seed production, wind-firmness, and timber quality. The second stage of the shelterwood, the final clearcut, should occur about 10 years after the seed cut. If you use progressive strips, work with sets of three—the first two being clearcut at 10-year intervals and the third one cut in two stages as previously described.

You may need to control associated trees before the final harvest if you want to obtain 50 to 80 percent white-cedar on good sites managed for timber or deer habitat. Kill undesirable trees (especially hardwoods) that reproduce by root suckers or stump sprouts at least 5 and preferably 10 years before reproduction cutting.

Rely on residual stems to reproduce a stand only if there are at least 600 stems per acre of relatively young (less than 50 years old) and healthy white-cedars remaining. Remove heavy slash that buries residual stems or seedbeds. Full-tree skidding in winter will remove most slash and is recommended where residual trees will be relied on for reproduction. Either full-tree skidding or burning may be used for slash disposal in clearcut strips.

Pests

White-cedar is relatively free of major insect and disease problems. Wind may cause breakage and uprooting, mainly along stand edges and in stands opened up by partial cutting. White-tailed deer and snowshoe hare commonly browse northern white-cedar so severely that a stand cannot become established. Overbrowsing may be minimized when regenerating stands if large patches (40 acres or more) are completely cleared. Roads, beaver dams, and pipelines that impede the normal movement of soil water will kill northern white-cedar.

OAK-HICKORY

Products and Uses

Oak is valued in furniture, flooring, paneling, ties, cooperage and fuelwood. Hickory, a wood of great strength, has much less market demand but is used for furniture, tool handles, high-strength specialty items, flooring, plywood, fuelwood, and charcoal.

The oak-hickory forest is home to many game animals such as white-tailed deer, turkey, gray and fox squirrels, and ruffed grouse. Raccoon, opossum, red fox, bobcat, skunk, and many birds also take advantage of this forest type.

Fig. 46. Range of northern red oak.

Growing Conditions

Northern red oak, white oak, bur oak, black oak, northern pin oak, and various hickories make up an oak-hickory forest. Associated species include red and sugar maple, black cherry, American basswood, black walnut, white pine, and white and green ash. If left undisturbed, an oak-hickory stand in the Upper Midwest will shift toward more shade-tolerant species.

Oaks grow best on north- and east-facing, gently sloping, lower slopes where soils are at least 36 inches deep. Medium-quality sites have moderately deep soils (20 to 36 inches) on upper and middle slopes facing north and east. Oaks survive but grow poorly on narrow ridgetops or south- and west-facing, steep, upper slopes where soil is less than 20 inches deep. Oaks survive better than most other tree species on dry sites, but they do not produce much merchantable timber on such sites. On the best sites there is fierce competition among tree species, so oaks are difficult to regenerate there.

Site index curves for northern red oak are shown in Appendix C-10.

Regeneration

Oaks commonly reproduce from acorns. Northern red oaks produce good acorn crops at 2- to 5-year intervals; however, seedlings are abundant only following years when there are excess acorns not damaged by acorn weevils or consumed by wildlife. Acorns usually are disseminated by blue jays, squirrels, and gravity. Best germination occurs in mineral soil under a light covering of leaves.

Oaks also reproduce from stump sprouts following a harvest. Sprouting frequency declines as tree diameter increases. Northern red oaks sprout more frequently than white oaks.

Northern red oak and white oak are intermediate in shade tolerance. In stands with a dense understory or overstory, there will be few oaks in the understory.

Oaks may live for several hundred years. However, for timber production in this region, stands on moderate to good sites can be regenerated when the oaks are 60 to 90 years old and trees average 18 to 24 inches in DBH. A stand also may be ready for harvest and regeneration if it is greatly understocked or most trees are poor quality or undesirable species.

Natural oak regeneration is most reliable where there is plenty of advance regeneration. The number of oak seedlings needed to successfully stock the next stand depends on seedling size prior to the harvest. The larger the seedlings, the more likely they are to survive to harvestable size. For example, you need 15,435 seedlings per acre if they are less than 1 foot tall. If they all are more than 4 feet tall, you need just 514 per acre.

Stands that are well stocked with advance regeneration and that have relatively little competition from undesirable understory trees, shrubs, or other vegetation may be clearcut and usually regenerated successfully. A clearcut should be at least 1/2 acre and preferably at least 2 acres in size; otherwise, shade from the surrounding timber will suppress oak seedling growth. For regeneration purposes there is no maximum size for clearcuts.

If there is a seed source present but few seedlings, the problem often is too much shade. Acorns may germinate and the seedlings may survive for 4 or 5 years beneath heavy shade, but advance oak regeneration will not accumulate over a long period.

To resolve this problem, start by reducing shade produced by the understory of shade tolerant hardwood trees, shrubs, or ferns. Cut or treat vegetation with herbicide, depending on the species to be controlled. If dense shade is produced by a high canopy, also remove 20 to 40 percent of the canopy in a shelterwood harvest, leaving species and individual trees that you want to provide seed for the next generation. Harvest carefully to avoid damaging residual timber.

Following a shelterwood harvest and understory removal, you may need to wait several years before clearcutting the stand to ensure that a satisfactory number of oak seedlings are present. You may need additional understory control if the oaks take more than 5 years to regenerate.

An alternative to the shelterwood harvest is to wait until there is a good acorn crop, then clearcut and disturb the soil after the acorns drop but before the ground freezes. Soil disturbance helps to bury the acorns and uproot competing vegetation.

Because of the risk and possible delay involved when relying on natural regeneration, you may want to plant seedlings. Planting enables you to supplement natural regeneration, to use genetically superior stock when it is available, and to choose the species. Seedlings may be planted immediately following a shelterwood cut or a clearcut. Before planting, control undesirable trees and shrubs by cutting, bulldozing, or treating with herbicide.

The best oak seedlings have a fibrous root system and a stem at least 3/8-inch in diameter. If large seedlings appear difficult to handle during the planting operation, just prior to planting clip the tops of the seedlings, and the roots if necessary, leaving each about 8 inches long. You may need to plant 200 to 800 seedlings per acre, depending on their size and the amount of advance reproduction already in the stand. Control weeds around the oak seedlings for 1 to 2 years. Herbicides are often effective and economical for weed control.

Intermediate Treatments

Control undesirable tree species that compete with crop trees when stand height averages at least 25 feet (10 to 20 years old). When growing trees for timber production, thin sprouts growing from a single stump to one or two dominant sprouts that have good form and are connected to the stump below or near the ground. Thin when sprouts are about 10 years old (2 to 3 inches in diameter). Oak stands managed for timber should be kept fairly dense until the bottom 20 to 25 feet of the stems are essentially free of live branches. This generally will occur when trees are 40 to 50 feet tall (30 to 45 years). At this stage thin stands to stimulate diameter growth of crop trees. Follow the crop-tree release guidelines in Chapter 5. Release no more than 100 crop trees per acre. As an alternative to the crop-tree release guidelines you can follow the stocking charts for upland central hardwoods in Appendix D-5.

Pruning will improve wood quality and may be needed if stand density is not sufficient to cause natural pruning. Follow guidelines in Chapter 5.

Pests

The red humped oakworm, two-lined chestnut borer, oak wilt, and shoestring root rot are significant pests on oaks.

Minimize chestnut borer damage by maintaining vigorous stands. Specifically, in upland stands with a site index less than 65 (Appendix C-10), maintain basal areas at less than 120 square feet per acre in stands with trees averaging 7 to 15 inches DBH, and at less than 100 square feet per acre in stands averaging more than 15 inches DBH. Avoid thinning for five years after a serious drought or defoliation.

To minimize oak wilt infections, do not thin or prune oaks from mid-April through mid-July, when fungal spores are present and can be transported by picnic beetles to fresh wounds. Dormant season operations are best because spores are not present and the trees are not susceptible to infection. Since oak wilt commonly spreads through root grafts between neighboring oaks, surround valuable oak stands in areas with a high oak wilt hazard with a 100-foot buffer of an alternate species. If trees become infected, harvest them before the following spring. Use a trenching machine or vibratory plow to break the root grafts through which the disease spreads. Trench placement and depth are critical. Consult a forester for advice before trenching. Left untreated, oak wilt will spread through the stand until it kills all red oaks. White and bur oaks are not commonly affected by oak wilt.

The red humped oakworm defoliates late in the year. It causes little growth loss, but weakens the trees and makes them more prone to shoestring root rot and two-lined chestnut borer. High value stands can be sprayed. No cultural control is available for oakworm, but maintaining stand vigor should help minimize damage.

RED (NORWAY) PINE

Products and Uses

Red (Norway) pine commonly is used for pulpwood to produce high-grade printing and wrapping papers. It also is used for lumber, veneer, pilings, poles, cabin logs, and posts. Red pine stands generally are considered poor habitat for game birds and animals, but old-growth trees are used as nesting sites by bald eagles and many songbirds.

Growing Conditions

Red pine often grows in relatively pure, even-aged stands. On drier sites it grows in pure stands or in mixture with jack pine, aspen, paper birch, and oaks. On moist sites it often grows in combination with eastern white pine, red maple, northern red oak, balsam fir, or white spruce. It is common on sandy soils where the site index ranges from 45 to 75, but grows best on well-drained (but not dry) sandy to loamy soils. Heavy, wet soils are poor sites for red pine.

Site index curves for red (Norway) pine are shown in Appendix C-11.

Regeneration

Because red pine is shade intolerant and good seed crops occur at 3- to 7-year intervals, clearcutting followed by planting is the most reliable regeneration method. A common spacing is 7 x 7 feet. Trees can be planted at wider spacings (up to 10 x 10 feet) if high survival is expected. Closer spacing reduces tree taper and branch size, promotes early crown closure, and suppresses competition, but also requires more frequent thinnings. If precommercial thinnings are not feasible, avoid close spacing.

The most common planting stock is 3-0 bare-root seedlings, but 2-0 seedlings sometimes are used. Containerized seedlings may be used to extend the planting season.

Site preparation prior to planting should reduce competition for light, water, and nutrients without causing any serious soil loss. On most sites shrubs should be controlled and mineral soil exposed. Mechanical equipment, herbicides, prescribed burning, or a combination may be used to prepare the site.

Intermediate Treatments

Cultural practices are needed to keep red pine crop trees free from overhead shade and to provide needed growing space. Red pine seedlings may need a complete release from shrubs and other low competition by the second or third growing season. Release plantations overtopped by hardwoods as soon as possible. In seedling stands (less than 2 inches average DBH) with more than 2,000 trees per acre, at least 100 potential crop trees should be given a minimum growing space of 25 square feet each. Dense sapling stands (2 to 5 inches average DBH) with 160 square feet or more of basal area per acre should be thinned to give 50 square feet of growing space per tree (7 x 7-foot spacing).

The stocking chart for red pine (Appendix D-6) provides guidelines for thinning pole and sawtimber stands. Pole stands (5 to 9 inches DBH) with greater than 140 square feet of basal area per acre should be thinned to about 90 square feet of basal area per acre. Small sawtimber stands (9 to 15 inches average DBH) grow well at densities around 120 square feet of basal area per acre. Large sawtimber (15 inches or larger average DBH) can be managed at densities of 150 or even 180 square feet of basal area per acre. As a general rule, remove less than half of the basal area in any one thinning, and during early thinnings, cut trees that are smaller, slower growing, and poorer quality than the stand average.

To produce high-quality sawtimber, prune crop trees to a height of 17 feet when they are poletimber-size.

Fig. 47. Range of red (Norway) pine.

Pests

Bark beetles are very serious pests, particularly in dense stands on sandy soils during drought years. They can be managed using the measures discussed for eastern white pine.

Scleroderris canker, red pine shoot blight, diplodia, root rots, butt rots, and needle blights may be important in some areas. The best control measures are to remove infected trees and maintain stand vigor through favorable growing conditions. Avoid establishing young stands beneath or near infected older pine trees.

Defoliating insects include sawflies and jack pine budworm. They can be controlled with insecticides where needed. The European pine shoot moth and the Zimmerman pine moth damage tips and buds, resulting in deformation of the main stem. Cultural controls are not effective, and insecticides have limited use against these insects. The Saratoga spittlebug, white grubs, and pine root collar weevils also injure or kill red pine. Spittlebug is controlled by removing sweetfern, grubs by killing sod, and root collar weevil by pruning lower branches and raking up needles near the tree's base.

Animal injury may be caused by deer, hare, rabbit, porcupine, or mice. Eliminate protective grass to decrease hare, rabbit, and mouse activity. Animal control or repellents may be necessary in other cases.

SPRUCE-FIR

Products and Uses

White spruce and balsam fir are used mainly for pulpwood and small sawtimber. Spruce-fir stands provide habitat for grouse, songbirds, white-tailed deer, moose, and a variety of small mammals.

Fig. 48. Range of white spruce.

Growing Conditions

White spruce often forms pure stands but may be associated with black spruce, balsam fir, aspen, and paper birch. Stands that develop following a major disturbance such as fire, insect attack, or clearcutting will be even-aged. Since white spruce and balsam fir are shade tolerant, uneven-aged stands will develop over time if no disturbances occur.

White spruce and balsam fir grow on a wide variety of soils, but white spruce does best on well-drained loams and clays. Balsam fir grows best on moist, well-drained sandy loam. The spruce-fir cover type commonly is found on well-drained lowlands where there is less competition from hardwoods.

Site index curves for white spruce are shown in Appendix C-12.

Regeneration

White spruce and balsam fir regularly produce good seed crops and often are regenerated by natural seeding. Almost any moist seedbed is adequate, but bare mineral soil with some shade is best. The spruce-fir type can be managed under the selection system, but a shelterwood cut or strip clearcut less than 6 chains wide is recommended to produce even-aged stands. These smaller cutting blocks, when surrounded by seed-producing trees, permit good seed dispersal and reduce windthrow of surrounding trees. They also produce an age-class mosaic less susceptible to a spruce budworm attack.

If natural spruce regeneration is desired, expose mineral soil. Full-tree skidding usually will expose enough mineral soil for natural seeding unless the ground is frozen.

When planting white spruce seedlings, first remove heavy slash concentrations. Plant 600 to 1,000 seedlings per acre. Transplant stock is preferred, but because of its high cost, 3-0 seedlings or container-grown stock often are used. Balsam fir is not often planted because of low market demand and relative ease of regeneration by natural seeding.

Intermediate Treatments

Beginning at age 25 to 30, thin the stand every 15 to 20 years, to a basal area of about 90 square feet per acre. Follow the stocking chart for even-aged spruce-balsam fir stands in Appendix D-7. White spruce may be grown on 80- to 120-year rotations for sawlogs. Harvest the balsam fir and aspen in the stand when they mature (age 60) to encourage spruce growth.

Pests

The spruce budworm is the major insect pest of the spruce-fir forest type. Budworm survives best on older trees. To minimize damage, manage fir on a 40- to 50-year rotation, keep large forest areas well diversified by age class, and maintain a high spruce and hardwood component. Insecticides may be warranted in areas where budworm attack is prolonged.

Heart rot and root rots are major diseases. They can be minimized by keeping budworm from killing the tops of trees (insecticides may be required) and avoiding scarring residual stems during intermediate cuttings.

Windthrow can be a serious condition, especially on wet, shallow soils. Minimize windthrow by maintaining a well-stocked, vigorous stand.

TAMARACK

Products and Uses

Tamarack (eastern larch) is used for pulp, poles, and lumber, although it has relatively minor economic importance. Red squirrel, snowshoe hare, and porcupine are found in tamarack stands. The tamarack is habitat for many songbirds and is critical habitat for the great gray owl and its small mammal prey species.

Growing Conditions

Tamarack is found in pure stands, but more commonly in mixed stands with black spruce, northern white-cedar, black ash, red maple, eastern white pine, or paper birch. Tamarack stands usually are even-aged.

Tamarack commonly grows on peatland where the organic soil or peat is more than 12 inches thick. It occurs on a wide range of peatlands, but is most characteristic of poor swamps where soil water is weakly enriched with mineral nutrients. The best sites are moist, well-drained loamy soils along streams, lakes, or swamps, and mineral soils with a shallow surface layer of organic matter. It grows well on upland sites, but is quickly eliminated by competition from more shade-tolerant species. Tamarack will not survive prolonged flooding.

Site index curves for tamarack are shown in Appendix C-13.

Regeneration

The regeneration system advised for tamarack is a combination of clearcut and seed-tree with natural seeding. Good seed years occur every 3 to 6 years starting when trees are about 40 years old. The best seedbed is a warm, moist mineral or organic soil with no brush, but a light cover of grass or other herbaceous vegetation. Hummocks of slow-growing sphagnum moss often make a good seedbed. Most seed falls within 200 feet of the seed tree.

Harvest strips should be oriented perpendicular to the wind and may be up to 200 feet wide. After clearcutting the first strip, wait about 10 years or until the area is well stocked with seedlings, then clearcut a second strip adjacent to the first and on the windward side. Again wait until regeneration is established, then use the seed-tree

Fig. 49. Range of tamarack.

method to cut the remaining strip. The seed-tree cut should leave about ten well-spaced dominant tamaracks per acre. Once the regeneration is established, harvest or kill seed trees.

You may need to prepare the site following a harvest to assure tamarack regeneration. Broadcast burn mixed species stands to remove slash (see Chapter 7). Since tamarack slash does not burn well, harvest pure tamarack stands by full-tree skidding to remove slash, then treat the brush with herbicides. Alternatively, you could pile and burn the slash or shear or chop the brush.

Tamarack seedlings need abundant light and constant moisture. Seedlings established under a fully stocked stand will not survive beyond the sixth year. Early seedling losses are caused by damping-off fungus, drought, flooding, inadequate light, and snowshoe hares. Given enough light, tamarack is one of the fastest growing conifers on upland sites.

Intermediate Treatments

Thinning is economically feasible only on good sites when the object is to produce poles or sawtimber. If a market exists for small products such as posts or pulpwood, make a commercial thinning as soon as the stand produces these products. Additional periodic thinnings are recommended up to 20 years before the end of the rotation. Each thinning should leave a basal area of 80 to 90 square feet per acre.

Pests

The larch sawfly is a serious insect pest that can kill tamarack after several years of defoliation. Chemical control may be required to manage sawfly populations. There is no effective cultural control for sawfly. Bark beetles can kill tamarack stressed by defoliation or competition in densely stocked stands. Tamarack also is susceptible to root and heart rots. Minimize rots by avoiding damage during intermediate cuttings. Porcupines can cause extensive damage by feeding on the bark of the main stem.

Suggested References

Bates, P. C., C. R. Blinn, and A. A. Alm. 1991. *Regenerating Quaking Aspen: Management Recommendations (NR-FO-5637).* University of Minnesota, Minnesota Extension Service, St. Paul, MN 55108. 8 pp.

Benzie, J. W. 1977. *Manager's Handbook for Jack Pine in the North Central States (General Technical Report NC-32).* USDA Forest Service, North Central Forest Experiment Station, 1992 Folwell Avenue, St. Paul, MN 55108. 18 pp.

Benzie, J. W. 1977. *Manager's Handbook for Red Pine in the North Central States (General Technical Report NC-33).* USDA Forest Service, North Central Forest Experiment Station, 1992 Folwell Avenue, St. Paul, MN 55108. 22 pp.

Brinkman, K. A. and E. I. Roe. 1975. *Quaking Aspen: Silvics & Management in the Lake States (Agriculture Handbook No. 486).* U.S. Government Printing Office, Washington, DC 20402. 52 pp.

Erdmann, G. G., T. R. Crow, R. M. Peterson, and C. D. Wilson. 1987. *Managing Black Ash in the Lake States (General Technical Report NC-115).* USDA Forest Service, North Central Forest Experiment Station, 1992 Folwell Avenue, St. Paul, MN 55108. 9 pp.

Frank, R. M. and J. C. Bjorkbom. 1973. *A Silvicultural Guide for Spruce-Fir in the Northeast (General Technical Report NE-6).* USDA Forest Service, Northeastern Forest Experiment Station, 5 Radnor Corporate Center, Suite 200, 100 Matsonford Road, Radnor, PA 19087. 29 pp.

Jacobs, R. D. and R. D. Wray. 1992. *Managing Oak in the Driftless Area (NR-BU-5900).* Minnesota Extension Service, University of Minnesota, St. Paul, MN 55108. 32 pp.

Johnston, W. F. 1977. *Manager's Handbook for Black Spruce in the Northcentral States (General Technical Report NC-34).* USDA Forest Service, North Central Forest Experiment Station, 1992 Folwell Avenue, St. Paul, MN 55108. 18 pp.

Johnston, W. F. 1977. *Manager's Handbook for Northern White-Cedar in the North Central States (General Technical Report NC-35).* USDA Forest Service, North Central Forest Experiment Station, 1992 Folwell Avenue, St. Paul, MN 55108. 18 pp.

Johnston, W. F. 1986. *Manager's Handbook for Balsam Fir in the North Central States (General Technical Report NC-111).* USDA Forest Service, North Central Forest Experiment Station, 1992 Folwell Avenue, St. Paul, MN 55108. 27 pp.

Lancaster, K. F. and W. B. Leak. 1978. *A Silvicultural Guide for White Pine in the Northeast (General Technical Report NE-41).* USDA Forest Service, Northeastern Forest Experiment Station, 5 Radnor Corporate Center, Suite 200, 100 Matsonford Road, Radnor, PA 19087. 13 pp.

Myers, C. C. and R. G. Buchman. 1984. *Manager's Handbook for Elm-Ash-Cottonwood in the North Central States (General Technical Report NC-98).* USDA Forest Service, North Central Forest Experiment Station, 1992 Folwell Avenue, St. Paul, MN 55108. 11 pp.

Ohmann, L. F., H. O. Batzer, R. R. Buech, D. C. Lothner, D. A. Perala, A. L. Schipper, Jr., and E. S. Verry. 1978. *Some Harvest Options and Their Consequences for the Aspen, Birch, and Associated Conifer Forest Types of the Lake States (General Technical Report NC-48).* USDA Forest Service, North Central Forest Experiment Station, 1992 Folwell Avenue, St. Paul, MN 55108. 34 pp.

Perala, D. A. 1977. *Manager's Handbook for Aspen in the North Central States (General Technical Report NC-36).* USDA Forest Service, North Central Forest Experiment Station, 1992 Folwell Avenue, St. Paul, MN 55108. 30 pp.

Safford, L. O. 1983. *Silvicultural Guide for Paper Birch in the Northeast (Research Paper NE-535).* USDA Forest Service, Northeastern Forest Experiment Station, 5 Radnor Corporate Center, Suite 200, 100 Matsonford Road, Radnor, PA 19087. 29 pp.

Schlesinger, R. C. and D. T. Funk. 1977. *Manager's Handbook for Black Walnut (General Technical Report NC-38).* USDA Forest Service, North Central Forest Experiment Station, 1992 Folwell Avenue, St. Paul, MN 55108. 22 pp.

Tubbs, Carl H. 1977. *Manager's Handbook for Northern Hardwoods in the North Central States (General Technical Report NC-39).* USDA Forest Service, North Central Forest Experiment Station, 1992 Folwell Avenue, St. Paul, MN 55108. 29 pp.

U.S. Department of Agriculture, Forest Service. 1983. *Silvicultural Systems of the Major Forest Types of the United States (Agricultural Handbook No. 445).* U.S. Government Printing Office, Washington, DC 20402. 191 pp.

U.S. Department of Agriculture, Forest Service. 1985. *Northern Hardwood Notes (Stock No. 001-001-0068-4).* U.S. Government Printing Office, Washington, DC 20402.

U.S. Department of Agriculture, Forest Service. 1990. *Silvics of North America, Volume 1 Conifers (Agricultural Handbook No. 654).* U.S. Government Printing Office, Washington, DC 20402. 675 pp.

U.S. Department of Agriculture, Forest Service. 1990. *Silvics of North America, Volume 2 Hardwoods (Agricultural Handbook No. 654).* U.S. Government Printing Office, Washington, DC 20402. 877 pp.

7

FOREST PROTECTION

During its long life, a tree may be subject to damage by animals, environmental conditions, insects, diseases, or fire. Consequently, woodlands should be thoroughly inspected at least annually and preferably more often. It is especially important to inspect stands during the growing season to detect serious problems.

It can be very costly to control damaging agents once they become established. For this reason, it is wise to prevent damage before it occurs.

You can minimize damage by:

- Matching tree species to sites where they grow best.
- Maintaining tree species diversity by either mixing tree species within a stand or growing several species in pure stands.
- Regulating stand density to encourage fast growth while maintaining relatively full stocking.
- Using pest-resistant planting stock when available.
- Pruning or thinning during the winter rather than the growing season.
- Avoiding wounding trees when operating heavy equipment or logging in the woodland.

Common sources of tree damage are described below. These descriptions are intended to alert you to the types of damage that may occur, their causes, and their potential severity. Chapter 6 briefly describes the most common pests associated with different forest types and their control or prevention. Contact your local forester to help you identify problems, assess damage, and design a control strategy for any serious problem. The forester also can help you plan stand management practices that will reduce future problems.

ANIMAL DAMAGE

Animals that frequently damage trees include birds, deer, small mammals, and livestock.

Birds

Yellow-bellied sapsucker damage appears as parallel rows of evenly spaced, shallow, 1/4-inch or larger holes around a tree stem. These holes reduce wood quality, but threaten tree health only when they completely girdle the stem. Holes from other woodpecker species are larger, more irregular, and much deeper. Sapsuckers feed on the sap that flows from the holes. Other types of woodpeckers seek insects that live beneath the bark surface or are attracted to the sap. Other birds, such as pine grosbeaks, eat buds and cause minor damage.

Bird damage usually is limited so no control measures are needed in woodlands. Most bird species (except English sparrows, European starlings, and pigeons) are protected by federal law. **A federal permit is needed to destroy these birds, even if they are pests.** Check with the U.S. Fish and Wildlife Service before attempting any control that may harm birds.

Physical measures that chase birds away or protect a tree may simply route birds to other areas where they may injure other trees. Some techniques that have been found useful in protecting valuable trees in a yard include netting, metal barriers, visual repellents (e.g., model owls and snakes), loud noise, sticky repellents, and bad-smelling or-tasting chemicals (e.g., mothballs).

Deer

Deer browse on succulent shoots or seedlings and rub their antlers on young trees. Browsing may seriously damage regeneration where there is a high deer density and relatively little natural browse or few agricultural crops available. Deer can be particularly damaging to plantations. Antler rubbing may kill small trees, but few trees are damaged in this manner so control is not warranted.

In most cases, the deer population can be controlled by hunting according to state regulations. High fences are effective deterrents, but are prohibitively expensive for forestry purposes. Electrical fences with high voltage and low impedance are more economical and in some cases may be justified to protect young, high-value plantations. Contact a forester or wildlife damage control expert for advice on fencing. Commercial chemical repellents sometimes are effective, but may need reapplication after a rain or wet snow. You also can minimize forest openings and brushy habitat, making your forest less attractive to deer, or make cutting blocks and regeneration areas as large as possible, thus providing more browse than the deer can consume.

Small Mammals

Rabbits, snowshoe hares, and mice can girdle or cut off young trees near the ground. Reduce damage by eliminating high grass and brush piles that provide cover for these animals.

Pocket gophers feed on roots and bark around the base of young trees and other plants, especially in sandy soils. The best control is to reduce their food by eliminating as much vegetation as possible in the plantation. In small areas gophers can be trapped. Where the population is high and the plantation is large, they can be controlled using a device that creates an artificial tunnel and drops poisoned grain into it. Take great care when using poisoned grain to prevent spills and accidental poisoning of nontarget animals.

Beavers commonly dam small streams and flood woodlands, killing trees in the area. They also fell trees near streams and lakes in order to eat bark and small branches. Generally the only way to solve a beaver problem is to trap and remove the animals. Contact a forester or wildlife conservation officer for information on trappers in your area.

Porcupines eat bark and may girdle and kill some trees, especially white and red pines. If damage is serious, they can be live-trapped or killed.

Livestock

High concentrations of cattle, horses, or other livestock compact the soil in a woodland, trample young seedlings and sprouts, damage roots, rub bark from stems, and eat or defoliate small trees. Heavy grazing and forest management are not compatible on the same site. Fence livestock out of the woodland if you expect to grow high-quality trees.

ENVIRONMENTAL DAMAGE

Trees may be damaged by airborne chemicals, machinery, soil-related problems, too much or too little water, and severe weather.

Airborne Chemicals

Airborne pollutants from industry, automobiles, fires, and other chemical sources can harm trees. Symptoms include defoliation; browning or yellowing leaf margins, intraveinal tissue, or leaf tips; and stunted foliage growth. Herbicides may cause distortion, curling, and browning margins, or leaf drop in deciduous trees and may cause conifer needles to turn yellow or brown, and succulent shoots to curl and deform. Trees often survive such damage, but their growth may be stunted or their shape deformed. Where air pollution could be a problem, plant resistant species and maintain well-thinned stands. Follow label directions and standard application guidelines to minimize chemical injuries. While most pesticide-induced tree damage is caused by herbicides, insecticides and fungicides also can damage trees, particularly when the tank mix is too strong.

Machinery

Logging equipment can knock over small trees, break branches, tear bark, or destroy roots near the surface. Much of this damage can be avoided by using careful and experienced equipment operators or by providing a physical barrier between the trees and the equipment.

Cultivation equipment used in plantations can be very damaging to tree roots if it cuts too deep. Set cultivation equipment as shallow as possible and keep it away from tree stems. A good rule of thumb is to maintain a no-till strip 1 foot wide for each inch of tree diameter. You may be able to reduce damage by using a herbicide rather than cultivation equipment.

Soil

Soil compaction cuts off water and carbon dioxide to tree roots. It may be indicated by dying leaves on mature trees and dying branches on younger trees, but most often reduced growth is the only sign. Soil compaction is a potential problem mainly on wet clay or silt soils. The best way to avoid it is to operate heavy equipment in a stand only when the soil is frozen or dry.

Changes in the soil level around a tree also will affect growth. Excavating soil and severing roots may lead to windthrow or root diseases; adding soil decreases air movement to the roots. Avoid either removing or adding soil near trees.

Mineral deficiency causes a wide range of symptoms, from foliage discoloration to reduced foliage size. A soil analysis will indicate which mineral is deficient. Upper Midwestern woodlands are rarely fertilized.

Water

Drought damage occurs when water loss through the leaves exceeds water uptake by the roots. It is most common on sandy and gravelly soils, which do not retain much water. It also occurs if the water table drops suddenly and remains low, depriving the trees of needed moisture. Drought symptoms include wilting, off-color foliage, and a general decline in vigor. Crowns of drought-stricken trees usually die from the top down. Insect and disease attacks often are triggered by drought.

To minimize drought damage, do not plant shallow-rooted species in areas of low rainfall or on sandy soils. Flooding for an extended time can suffocate roots and kill trees. The extent of damage depends on species (see Table 9), time of year, tree size, and tree vigor. Flooding for short periods in winter or early spring is seldom a serious problem, but flooding during the growing season can kill seedlings in just a few days. Completely submerged seedlings will be killed more quickly than trees that have their crowns above the water. In flood-prone areas such as river flats, plant only tree species that tolerate flooding.

Table 9. Relative tolerance of tree seedlings of selected species to flooding during the growing season.

HIGH	MODERATE	LOW
American elm	balsam fir	bigtooth aspen
black ash	basswood	bitternut hickory
eastern cottonwood	black spruce	butternut
green ash	black walnut	northern red oak
silver maple	bur oak	red pine
willow	eastern white pine	sugar maple
	jack pine	white ash
	northern white-cedar	white birch
	quaking aspen	white oak
	shagbark hickory	white spruce
	swamp white oak	
	tamarack	
	yellow birch	

Weather

High temperatures and drying winds cause rapid water loss from tree leaves. Water loss causes leaf margins to turn brown and leaves to fall prematurely. Do not plant susceptible trees (e.g., sugar maple) in locations exposed to strong sunlight or wind.

Early fall frosts or extremely cold weather shortly after leaf fall can injure succulent twigs and buds. Late spring frosts can kill buds that have begun growing. Trees usually survive, but growth, stem quality, and vigor can be dramatically reduced. To avoid freeze damage, plant trees from seed sources that originate no more than 200 miles south of the planting site. (Near the northern edge of a tree species' range, do not plant trees from seed sources more than 50 miles south of the planting site.) If your planting site is in a low-lying, frost-prone area, select a species that breaks bud relatively late in the spring to avoid frost damage. For example, black spruce breaks bud about ten days after white spruce.

Winter sunscald occurs in early spring when the sun heats and activates tissue during the day and then freezing temperatures kill the active cells at night. The injury appears as peeling bark over an elongated canker on the south to southwest side of the tree stem. Thin-barked trees such as young maples are most susceptible. The tree rarely is killed, but diameter growth is reduced, decay-causing organisms often enter the tree, and logs become degraded or ruined. In hardwood stands, avoid overthinning or pruning that might expose tree stems to too much direct sun, especially if the trees are growing near the northern limits of their range.

Winterburn and winter drying are caused by warm spring winds that dry the foliage while the roots are still frozen in the ground and cannot replace the lost moisture. This damage is common on most conifer species and may be recognized by the reddening and browning of needles in the spring. Trees usually survive, but massive defoliation may occur on trees located along exposed plantation borders or on trees along highways where road salt exacerbates the problem.

Strong winds can topple trees or break their branches. To prevent wind damage, leave a dense row of trees with intact lower branches around the perimeter of woodlands. When thinning old, dense tree stands, progress slowly over a period of years to minimize windthrow.

Hail can break off buds and branches and shred tree foliage. There is no prevention or control.

Lightning can split trees, cause spiral cracks in the trunk, shatter limbs, and start fires. Trees with crowns high above the general canopy level are most subject to lightning strikes. There is no control for lightning other than to remove high-risk trees before damage occurs. Once a conifer is struck by lightning, remove it quickly before bark beetles attack it and build up a population that can spread to other trees.

INSECT DAMAGE

Each kind of insect affects very specific tree parts. Insect damage is categorized here by the tree part affected by the insects. Strategies for minimizing insect damage are discussed in Chapter 6 for the more serious pests.

Defoliating Insects

Defoliating insects remove all or part of a tree's foliage. They weaken the tree by lowering its capacity to produce starch and sugars.

Foliage damage takes many forms. Insects that remove only the softer leaf tissue and leave the network of veins are called **skeletonizers. Leaf miners** bore into and eat the tissue between the upper and lower surfaces of the leaf. **Window feeders** eat one leaf surface leaving the other intact. **Case bearers** and **bag makers** construct and live inside individual movable cases made of webbing and foliage parts. **Needle tiers** and **leaf rollers** encase, fold, roll, or tie adjacent leaves and needles together with webbing. **Webworms** or **tent caterpillars** make and live within conspicuous webbed tents. Other insects eat the entire leaf or needle.

Sapsucking Insects

Sapsucking insects injure trees by removing tree fluids. They usually are not serious pests in woodlands. However, heavy attacks lower a tree's energy reserves and may lead to a secondary pest problem. The general symptoms of sapsucking injury are loss of vigor, deformed leaves or plant parts, yellowed leaves, or dead branches. Galls (abnormal tissue growths) also may form.

The destructive stage of the insect usually is required for precise identification, but sometimes the presence of feeding punctures, sooty mold, eggs, fine webbing, etc., may suffice. The Saratoga spittlebug has been a serious pest in red pine plantations. Several scale insects, such as the pine tortoise scale and the pine needle scale, are important in midwestern forests. Aphids, midges, and mites are other sapsucking organisms that may affect your trees.

Bud, Twig, and Seedling Damaging Insects

Most of these insects deform trees. The white pine weevil is one of the most damaging insects in conifer plantations (notably pine) in the Midwest. White pine is the favored host, but other pines and spruces also are attacked. The weevil larvae feed just below the terminal bud and cause forking and crooking, especially in open-grown trees from 2 to 20 feet tall. The new growth elongates slightly before dying because of the larval borings. The current year's shoot is not attacked, but commonly wilts into a "shepherd's crook." Another damaging weevil, the pales weevil, eats the bark on young seedlings.

Bark-Boring Insects

The succulent and nutritious inner bark on the main stem and large branches attracts many insects, most notably the bark beetles. They tunnel beneath the bark and can girdle a tree, thus preventing the normal movement of sugar and water. They hasten the death of weakened trees, attack apparently healthy trees during population explosions and drought, and lower lumber value. They also can introduce disease organisms such as Dutch elm disease fungus and blue stain fungus.

The pine engraver is the most common bark beetle in the Lake States pine stands. It mass-attacks healthy trees during drought. Flat-headed inner bark borers such as the bronze birch borer, two-lined chestnut borer, some weevils, and round-headed borers feed on the inner bark. They rarely kill or damage more than a few trees unless the trees are severely stressed by drought or defoliation.

Bark borers are difficult to control with contact insecticides because they are sheltered beneath the bark. Systemic insecticides also have little effect because these insects disrupt water movement in the tree. Cultural practices such as thinning that maintain tree vigor provide good protection.

Wood-Boring Insects

Wood-boring insects attack very low-vigor or recently killed trees and rarely are a problem in vigorous stands. While common, they rarely cause a tree's demise. They feed for several weeks in the bark before boring into the wood. Flat-headed wood borers, round-headed borers, horntails, powder-post beetles, ambrosia beetles, and ants are wood-boring insects of common concern. Problems with chemical controls are the same as with bark borers.

Root-Feeding Insects

Root-feeding insects are mostly a problem in nurseries or in young plantations where sod is well-established. They disrupt absorption and movement of water and nutrients. Root maggots, cutworms, root-bark beetles, white grubs, and root-collar weevils are examples of root-feeding insects. In plantations, control the sod before planting trees. Killing sod after planting may cause these insects to concentrate their attack on tree roots.

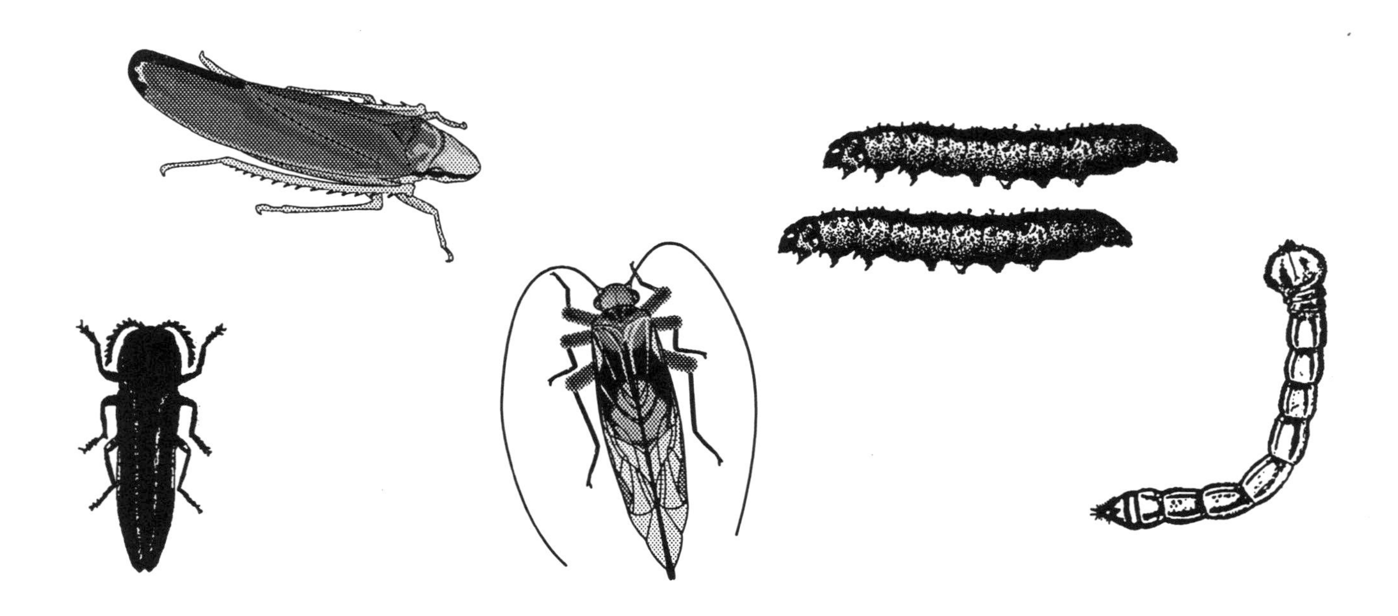

Cone and Seed Destroying Insects

Beetles, moths, and flies may destroy cones and seeds. Usually they deposit eggs in a seed or cone. The developing larvae then eat and destroy the seed. The red pine and white pine cone beetles, red pine cone worm, spruce cone worm, acorn weevil, and walnut weevil are some of the common seed-destroying insects. Insecticides can help control these insects; however, they are justified only in woodlands used as seed production areas or in oak stands where acorn production is critical for wildlife. Few insecticides are registered for this use. Most are very toxic and require application by a licensed commercial applicator.

DISEASE DAMAGE

Diseases are categorized here by the tree part they most commonly affect. Strategies for minimizing damage from the more serious diseases are discussed in Chapter 6.

Foliage Diseases

Foliage diseases my cause conifer needles to turn yellow or brown or drop prematurely. Hardwood leaves may develop yellow, brown, or black spots. These diseases weaken trees by reducing the ability of leaves to produce plant food. Brown spot disease affects only red and Scotch pines and is typically confined to the lower half of the tree. Some other typical foliage diseases are rhizosphera needle-cast of spruce, pine needle rust on conifers, and anthracnose and leaf spot on hardwoods.

When the leaves of a hardwood turn yellow (or brown) and droop, a wilt disease may be present. These symptoms commonly occur when a fungus blocks a tree's water-carrying vessels. Oak wilt and Dutch elm disease, verticillium, dothiorella, and phloem necrosis are typical wilt diseases. Oak wilt and Dutch elm disease are serious problems in woodlands and often spread to adjacent trees through root grafts.

Sooty mold is a black powdery fungus that lives on the honeydew exuded by aphids or scale insects. Powdery mildew is a white fungus that covers leaf surfaces. They damage trees and shrubs by blocking sunlight needed for photosynthesis. Their damage is a minor problem in woodlands, but may be serious on some ornamental trees and shrubs.

Abnormal growth, including leaf curling; gall formation on leaves, twigs and fruits; and witches' brooms (excessively dense branch and twig growth) are the result of high concentrations of plant growth-regulating compounds caused by insects, herbicides, or disease organisms. These conditions are rarely serious by themselves.

Stem and Branch Diseases

Cankers are dead areas on stems that are symptoms of diseases such as nectria canker on maple, hypoxylon canker on aspen, and scleroderris canker on pines. Affected areas may be irregular, sunken, flattened, or swollen. They may crack open and enlarge each year until they completely girdle the stem, killing the tree above the canker.

Rust diseases on pines may cause stem cankers and turn the foliage yellow before killing the tree. White pine blister rust is the most important rust in the Midwest. This stem disease requires gooseberry (currant) as an alternate host.

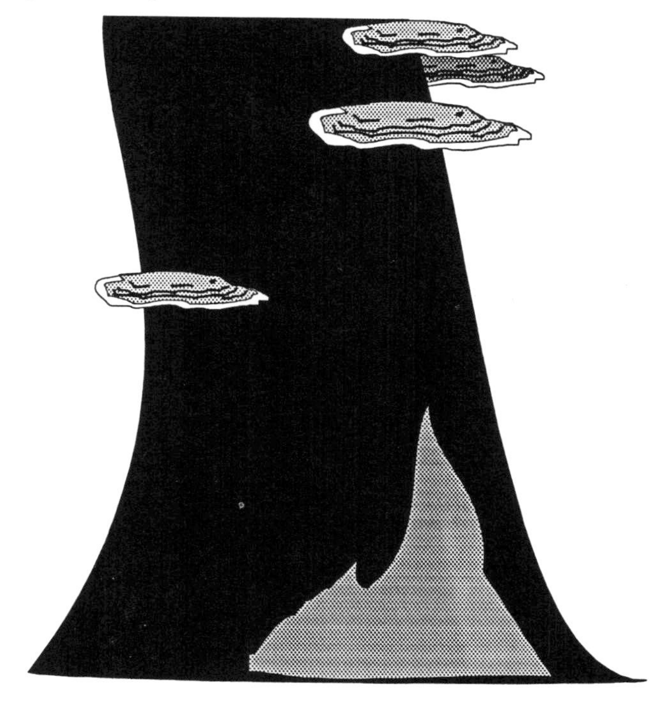

Dwarf-mistletoe is a parasitic plant causing a problem in the Lake States. It grows on limbs and small branches and may stunt, deform, or kill conifers. Its visible growth is less than 1 inch long and may be either single-stemmed or branched and yellow, brown, or olive green in color. Black spruce stands are the most common host for dwarf-mistletoe in the Lake States. Control includes destroying all trees in a cutting area as well as any infected trees within 60 feet and burning slash.

All trees are susceptible to wood rot. The most obvious signs are fruiting bodies (conks, mushrooms, etc.) that appear after the rot has been active for several years. Decayed wood may appear water-soaked and spongy or dry and crumbly. It usually is discolored. Many decay organisms enter through wounds in the stem or roots. These rots do not kill trees, but they can destroy the commercial value of the wood. Tree stems with rot are more easily broken by the wind and so can be hazardous.

Root Diseases

Root rot causes decline in tree vigor over weeks, months, or years. Twigs and branches die back; leaves appear small and yellowed and may drop or wilt in hot weather. Since the root system is damaged, infected trees do not respond normally to water or fertilizer and are susceptible to windthrow. Root rot also may cause stem decay.

FIRE DAMAGE

Wildfires can cause great damage to woodlands. They may weaken or kill trees, cause wounds where insects and diseases can enter, increase soil erosion, and reduce soil fertility, wildlife habitat, and recreational quality. Fire also can be used constructively to manage forest vegetation.

Forest fires are classified as surface, crown, or ground fires based on their manner of spread.

Most forest fires in the Midwest are surface fires. They burn only the litter and other small fuels on the forest floor. They may scar the bases of large trees and kill small trees.

Crown fires usually start as surface fires that reach into the canopy with the help of dry winds and fuel ladders. They occur most often in conifer stands and are very damaging and difficult to control. An intense crown fire will produce showers of sparks and glowing embers that easily jump firebreaks and set additional fires well in advance of the leading edge. Although conifer crowns frequently catch fire, true crown fires that spread through the air from one crown to the next are fairly rare.

A ground fire burns and smolders below the surface sometimes going undetected for days or weeks. It consumes soil high in organic matter including dried peat and thick litter. It produces enough heat to kill most of the trees in its path by cooking their root systems. Such a fire may cross firebreaks through roots and dry organic matter. Ground fires are very difficult to control, but are likely to occur only in dry years.

Few woodland owners can afford their own fire suppression equipment. Instead, most rely on state and local agencies to control fires. While these organizations respond quickly, there may be some delay before the fire is reported and crews arrive on the scene. For this reason you need to maintain your land so wildfires cause minimal damage before they are suppressed. The following practices will help:

- Maintain a cleared firebreak around your woodland. A firebreak might consist of a rough bulldozed road with a bare mineral soil surface that can be driven by a four-wheel drive fire truck. Although such a firebreak may stop a surface fire, it is more likely to be a good starting place for a fire suppression crew to build a fire line.
- If the woodland is more than 20 acres, consider establishing a trail/road system within it to provide access to all areas and break it into smaller, more defensible units. The road system also may provide access for other management activities or recreation.
- Create a pond. This will provide water for fire suppression and may benefit wildlife.
- Thin and prune pine and spruce-fir stands to keep stands from building fuel ladders that permit a surface fire to climb into the tree crowns.

- Create buffer strips of hardwoods around conifers for added protection. Hardwood stands are less flammable than conifer stands and also may diversify wildlife habitat.
- After timber harvests, lop or chip slash so that it lies close to the ground and decays quickly. You can also pile and burn slash, but take care not to start a wildfire.
- Cooperate with adjacent landowners in designing and establishing fire prevention measures.
- Place fire prevention and suppression clauses in logging contracts.

Controlled burns are fires set intentionally under specific fuel and weather conditions. They can be used in established stands to change the quantity and species of understory vegetation, thus enhancing the growth of overstory trees or benefiting wildlife. They can be used to reduce fuel loads that contribute to wildfire hazard. On recently harvested sites they are used to kill or set back the growth of undesirable trees and shrubs and to eliminate woody debris that hinders access for planting trees or that may harbor insect and disease pests.

The soil chemistry and physical processes on a burn site change temporarily, but will return to normal. However, a poorly planned or improperly controlled burn may kill crop trees. Consult a forester in regard to firebreak placement, weather requirements, tools needed, legal liabilities, and other important aspects. A controlled burn always poses some risk, but it should remain an option if it fits into your management plan.

Suggested References

U.S. Department of Agriculture, Forest Service. 1985. *Insects of Eastern Forests (Misc. Pub. 1426).* U.S. Government Printing Office, Washington, DC 20402. 608 pp.

Timm, R. M. (editor). 1983. *Prevention and Control of Wildlife Damage.* Great Plains Wildlife Council. Nebraska Cooperative Extension Service, Lincoln, NE 68583.

8

MARKETING TIMBER[1]

This chapter outlines procedures for selecting trees to harvest, obtaining bids, preparing a timber sale contract, and administering a sale. By working with a forester and following the steps recommended here, you can receive a fair price for your timber and meet your other woodland management objectives.

WHY HARVEST TIMBER?

There are several reasons why you may decide to harvest timber. First, harvesting is an important management tool to improve the health and vigor of the forest, promote natural regeneration, control stand density, release an established understory from undesirable overstory trees, develop wildlife habitat, alter species composition, establish planting areas, create vistas, and clear trails.

Second, the timber may have considerable value and can be managed to produce periodic or emergency income.

Third, timber may be harvested to salvage its value following damage by ice or snowstorms, high winds, fire, insects, or diseases.

Fourth, timber may be harvested in order to clear the land for other purposes.

STEPS IN MARKETING TIMBER

Follow these steps when marketing timber:

1. Select trees to harvest.
2. Determine timber worth.
3. Determine what method you will use to sell timber.
4. Determine what method you will use to select a buyer.
5. Determine how your timber will be priced.
6. Advertise your timber sale.
7. Develop a written contract with a buyer.
8. Inspect the harvest operation.

[1]Information in this chapter was adapted from Blinn, C. R. and L. T. Hendricks. 1991. *Marketing Timber from the Private Woodland (NR-BU-2723)*. University of Minnesota, Minnesota Extension Service, St. Paul, MN. 55108. 20 pp.

SELECTING TREES TO HARVEST

Select the trees to be harvested with advice from a forester to ensure that the harvest satisfies your management objectives and maintains the woodland in a vigorous and productive condition. For example, the harvest could range from a light thinning that stimulates growth of residual trees to a selection cut, clearcut, or shelterwood cut aimed at harvesting mature timber and regenerating a new stand. The type and amount of harvesting depends on your objectives and on stand conditions. For additional guidelines on selecting trees to harvest, refer to Chapter 6.

Clearly mark trees to be harvested so they can be identified easily by the logger. If scattered, individual trees are to be cut throughout the woodlot, as in a single-tree or group selection system, mark each with a paint spot at about chest height. Place paint spots on the same side of all marked trees (e.g., the north, south, east, or west) or mark the trees so they are visible from a main trail or road. Where the timber is valuable and there is a risk that the logger might harvest unmarked trees, place a second paint spot at ground level. This second spot should remain after logging to serve as a check that only marked trees were harvested.

If all trees in an area are to be harvested, as in a clearcut, mark only the boundary trees and instruct the logger to leave the marked trees uncut. If only scattered, individual trees must be left after harvesting, as in a seed tree harvest, mark only the leave trees and instruct the logger to cut all other trees.

After selecting the trees to be harvested, estimate the wood volume or number of products that will be cut by species. Products commonly produced in a timber sale include sawlogs, veneer logs, pulpwood, fuelwood, posts, or poles. Local mills or buyers will determine the specifications for each product they purchase.

Information about measuring wood volumes can be found in Chapter 2.

DETERMINING TIMBER WORTH

Timber is an unusual commodity in that it has no pre-established price. Instead, the price is whatever the buyer and seller agree to and is influenced by many factors, including:

1. **Tree species.** Wood from some species is more valuable than wood from other species.
2. **Tree size.** Large diameter trees have more usable volume and clear wood than small trees and are of greater per unit value.
3. **Tree quality.** Trees with fewer butt log defects (e.g., branch scars, decay, and embedded wire) have higher quality, more valuable wood.
4. **Sale volume.** On large sales, fixed logging costs can be spread over larger volumes, so the buyer can pay more per unit volume for the timber.
5. **Distance to market.** The closer the woodlot is to the mill, the lower the hauling costs.
6. **Site accessibility.** The ease with which the timber tract can be reached affects road construction costs.
7. **Logging difficulty.** Steepness of terrain and soil moisture conditions affect the equipment that can be used and speed of harvesting.
8. **Market conditions.** Poor markets mean lower timber prices.
9. **Mill's log inventory.** Buyers often pay more for logs when their inventories are low to ensure continued mill operation.
10. **Your restrictions on harvesting and skidding techniques.** Restrictions that protect the site and residual trees tend to increase logging costs.

A forester can estimate the expected value of a particular sale. However, different buyers may offer substantially different prices for the same timber, depending on their own particular costs and markets. To receive the highest value, contact several potential buyers when you have timber for sale.

METHODS OF SELLING TIMBER

You can either harvest your own timber and sell the logs or you can sell stumpage—standing trees—and allow the buyer to cut and haul the logs away.

Harvesting Your Own Timber

If you deliver harvested material to the roadside or the mill, you may receive higher income from the sale as a result of the investment of labor. Don't undertake such an operation, however, unless you have experience in timber harvesting.

Logging is hard, dangerous work that requires special skills and knowledge. If you hire labor, you will need to obtain liability insurance, pay workers' compensation, and complete special income tax forms. Some types of logging may require special equipment. Attempting to log with agricultural machinery not properly adapted for woodland use can result in extensive damage to the equipment. Logging often requires more time than is initially estimated and so may interfere with other activities and responsibilities.

Improper cutting, handling, or transporting of high-value logs can destroy a great deal of their value. For this reason, it's best to sell high-value timber such as black walnut trees containing veneer-quality logs as stumpage.

Basic instructions for harvesting timber can be found in Chapter 9. If you have the time, skills, and experience and wish to harvest and transport the products, keep the following points in mind:

1. Locate a market for your products before you harvest them. Obtain a written contract from the buyer.
2. Know your buyer's specifications for product size, quantity, and quality.
3. Know your legal responsibilities for workers' compensation, minimum wage, social security, state and federal tax, Occupational Safety and Health Act requirements, etc. These points are particularly important if you employ other people.
4. Use safe and efficient equipment.
5. Practice all appropriate safety precautions and procedures.

Selling Stumpage

Because of the many problems associated with harvesting your own timber, you generally are advised to sell stumpage. Then the buyer is responsible for harvesting your timber, employing people, obtaining machinery and equipment, and fulfilling all legal obligations associated with operating a business. Because this is the preferred alternative, the rest of this chapter assumes you are selling stumpage.

Pricing Stumpage

There are two general types of stumpage sales based on how the timber is priced. These types are the lump sum sale and sale-by-unit.

Lump Sum Sale

In a lump sum sale, you normally receive a single payment for the trees designated for sale. Alternatively, you may require a down payment of one-fourth to one-third of the sale price when the contract is signed and payment of the balance before harvesting begins. Payment is based on an estimate of the timber volume available in the sale area and not on the actual volume harvested. Lump sum sale values depend heavily on the accuracy of the timber inventory used to estimate the volume and quality of timber for sale. Lump sum sales may be appropriate if there is no convenient and reliable method for measuring the volume of cut logs.

Sale-by-Unit

In a sale-by-unit arrangement (also called sale-by-scale), you are paid a certain amount for each unit (thousand board feet, cord, post, ton, etc.) of product cut. A sale-by-unit requires that

Selling Timber by Weight

Timber sales by weight are increasing in frequency, especially for pulpwood in the Lake States. Under this method the trees are harvested and bolts are hauled to a mill. There the log truck is weighed both loaded and then unloaded, the difference being the weight of the wood. As a basis for payment, wood weight most accurately reflects the amount of usable wood when the trees are of a single species and logs are weighed when freshly cut (Table 10).

In some cases weight is converted to cordwood volume and the payment is based on volume rather than the actual weight. This conversion may lead to inaccurate measurement because wood weight varies by tree species, time of year, and time since the trees were felled.

Table 10. Approximate green weight of a cord of wood and bark by species.

SPECIES		APPROXIMATE GREEN WEIGHT PER CORD
aspen	SUMMER (MAY - OCT.)	4,300 lb. (2.15 tons)
	WINTER (NOV. - APR.)	4,500 lb. (2.25 tons)
balsam fir, red (Norway) pine		4,700 lb. (2.35 tons)
balsam poplar		4,800 lb. (2.40 tons)
Dense Northern Hardwoods (oak, hard maple, elm, walnut, yellow birch)		5,500 lb. (2.75 tons)
Lighter Northern Hardwoods (black and white ash, paper birch, soft maple, willow, cottonwood, hackberry)		5,000 lb. (2.50 tons)
jack pine, basswood		4,600 lb. (2.30 tons)
northern white-cedar		2,900 lb. (1.45 tons)
spruce		4,200 lb. (2.10 tons)
white pine		4,400 lb. (2.20 tons)

SOURCE: Minnesota Department of Natural Resources. 1985. *Scaling Manual.* Division of Forestry, 500 Lafayette Road, St. Paul, MN 55155.

someone measure the products harvested (a process called scaling). Products can be scaled by the landowner, by a professional forester, by the buyer, or by a receiving mill. You and the buyer need to determine who will scale the products based on who can be trusted to provide the most accurate information at a reasonable cost.

Although final payment is based on the actual volume harvested, some landowners ask for a down payment before the harvest of at least one-fourth of the estimated total value. Some landowners also request additional payments during the harvest, with these payments equal to the estimated value of the next cutting block to be harvested. Payment is adjusted at the end of the harvest to compensate for overpayment or underpayment.

SELECTING A BUYER

You can determine the sale price and buyer for stumpage through a single offer, an oral auction, or written sealed bids.

Single Offer

One option is to negotiate the sale price with a single buyer. This procedure often produces a price well below what the timber is worth, because the buyer has no competition and the seller often is uninformed about the timber's value. However, the single offer may be the best method for you if:

- You have only a small amount of timber or poor quality timber to sell, so there may be only one buyer interested in the sale.
- Markets for the species and products for sale are so poor that few buyers would be interested.
- You know and want to work with a particular buyer who has a good reputation.

Oral Auction

A second option is to invite several buyers to inspect your timber and, at a given time and place, bid for it at an oral auction. To attract several bidders and create competition, you need to hold the auction at a convenient time and location. Auctions are most appropriate for high-value sales or when several timber tracts can be auctioned at one time, thus attracting several buyers.

Sealed Bids

A third option is to notify several potential buyers about the timber you have for sale, give them time to inspect your timber (usually four to six weeks), and request that they submit written sealed bids. Written sealed bids produce the best results for private woodland owners in most situations.

Open bids at a specified time and place. Select the highest bidder unless you have other information that influences your decision. To be fair to all bidders, no further price negotiations should take place after bids are opened, and unsuccessful bidders should be notified that the timber has been sold.

Exercise caution in selecting new operators or operators who have not previously logged in your area. Most buyers perform satisfactorily when all the trees in an area are to be cut. However, only experienced and careful buyers should be selected for a thinning or selection harvest in which valuable trees will be left standing. A forester may offer advice about the desirability of selecting a particular buyer.

Before making a final selection, you may want to ask the potential buyer for the names of a few woodland owners with timber similar to yours for whom the buyer has harvested timber. Call one or more of those owners and ask about the logging job that was done. With their permission you also could visit one of the harvest areas to look at the results.

ADVERTISING YOUR SALE

Foresters usually can provide a list of timber buyers. The most effective way to notify potential buyers about your timber sale is to send them timber sale notices. If you are unable to assemble a list of buyers or have special products to sell, you may want additional advertising for your timber sale. Place a brief advertisement in the newspaper directing interested buyers to contact you for a complete description of the sale. Newspaper advertisement may be particularly useful for locating firewood cutters. Some firewood cutters do not harvest other products and may not appear on a list of local timber buyers.

A timber sale notice should include the basic information that will later be part of the timber sale contract including:

- Seller's name and address.
- Location of the timber for sale (legal description and directions).
- Description of the timber to be sold (volume; method used to estimate volume; tree species, size, and quality).
- Type of bid you are seeking (lump sum or sale-by-unit) and whether you will choose a buyer according to written sealed bids or an oral auction.
- Time period and procedure for inspecting the timber. (Allow at least one month for buyers to inspect the timber.)
- Date, time, and place written sealed bids will be opened or oral auction will be conducted.
- Whether a deposit (usually 10 percent or more of the bid price) binding the offer must be paid when the contract is signed.
- When payment is to be made. (In a lump sum sale, ask for full payment before the start of harvesting. If this is not possible, negotiate a definite payment schedule that calls for

specific percentages at specified dates. In a sale-by-unit situation, negotiate a definite cutting and payment timetable with the buyer.)

- Any major conditions or limitations on the sale, such as a harvesting deadline, who has cutting rights to tops that could be sold as firewood, method of slash disposal, restrictions on access to the area, or conditions when loggers cannot operate (e.g., when the area is excessively wet). (Note that excessive restrictions on buyers may result in reduced bid prices or fewer bidders.)
- The requirement of a performance bond. (A performance bond is an amount of money over and above the sale price, usually 10 percent, posted by the buyer and held in escrow by the seller. Its purpose is to ensure that the buyer abides by and fulfills all terms of the contract. It should be returned to the buyer when all contract conditions have been met.)
- Statement that the logger will be expected to carry workers' compensation insurance and liability insurance.
- Method recommended for scaling products.
- When the successful bidder will be notified (usually within seven days after bids are opened) and how much time the buyer has to sign the contract and provide the down payment after being notified of an acceptable bid (usually ten days).
- Statement indicating you have the right to reject any or all bids.

PREPARING A CONTRACT

Prepare a signed written contract with the buyer to reduce the possibility of misunderstandings and disagreements and to provide each party with legal assurance that the other will abide by the terms of the sale. The contract does not have to be a complicated document, but it should indicate what you and the buyer have agreed to with respect to the sale. The contract contents should be similar to those listed above for the timber sale notice. A sample contract is included at the end of this chapter. Your forester also may have a sample contract.

INSPECTING THE HARVEST OPERATION

Before harvesting begins, review the timber sale contract with the buyer. Either you or your forester should visit the site with the logger to point out the sale boundaries, discuss the location of log landings and roads, and point out any hazards or areas that require special protection during logging. Once harvesting begins, you or your forester should visit the area frequently to make sure the harvest is proceeding according to terms of the contract and to discuss questions that might arise. Frequent visits will help you become familiar with timber harvesting operations and help you plan future timber sales. Keep in mind, however, that logging is a dangerous activity. Do not endanger yourself or the loggers by getting too close to an active operation.

If you observe any problems while checking the harvest operation, simple suggestions to the buyer usually will resolve them unless a flagrant violation is observed. Deal directly with the buyer or the buyer's designated representative. Do not complain or make suggestions to other individuals on the job.

When the job is completed and all provisions of the contract have been fulfilled, write a letter releasing the buyer from the contract and return the performance bond if one was posted.

Suggested Reference

Blinn, C. R. and L. T. Hendricks. 1991. *Marketing Timber from the Private Woodland (NR-BU-2723).* University of Minnesota, Minnesota Extension Service, Room 3 Coffey Hall, St. Paul, MN 55108. 20 pp.

Sample Timber Sale Contract

You are encouraged to contact an attorney for help in designing a contract that meets your specific needs.

This agreement is made and entered into between the parties below hereinafter called the SELLER and the PURCHASER.

Seller Name: ______________________________ Telephone: ______________

Seller Address: ____

Purchaser Name: ____________________________ Telephone: ______________

Purchaser Address: ____

SECTION I

The Seller agrees to sell and the Purchaser agrees to buy, under the terms and conditions hereinafter stated, all the timber marked or designated by the Seller on certain lands held by the Seller and described as follows: __________ acres in Section _____ ,Township _________ , Range __________ , in _______________ County, State of _________________. Timber to be harvested is marked or designated as follows: [Describe cutting blocks and how timber is marked].

SECTION II

The Purchaser and Seller hereby agree to the following payment schedule: [Insert Option A or B]

Option A. Lump Sum Sale:

The Purchaser agrees to pay the Seller [25% of sale value] dollars ($_______) when the contract is signed, and [75% of sale value] dollars ($_______) before any timber harvesting activity begins as compensation for timber harvested. The Purchaser also agrees to pay the Seller [10% of total lump sum sale price] dollars ($_______) when the contract is signed as a refundable deposit to guarantee performance of Sections III and IV of this contract.

Option B. Sale-by-Unit:

The Purchaser agrees to pay the Seller an installment in advance of cutting. The first installment of [25% of estimated sale value] dollars ($_______) shall be paid when the contract is signed and subsequent installments shall be paid before harvesting begins in the next designated cutting block and in an amount equal to the estimated value of standing timber in the next designated cutting block. The volume of timber actually harvested will be measured [Location where timber will be measured, who will measure timber, when timber will be measured] (Note: If the consuming mill is designated as the official measurer of the timber volume, all scale receipts shall be provided to the seller.)

Within thirty days after the total volume of timber harvested is finally determined, the Purchaser agrees to pay or the Seller agrees to refund any difference in value from the original payment based on the actual scale at the unit prices specified below.

SPECIES	PRODUCTS	ESTIMATED VOLUME	UNITS	UNIT PRICE

The Purchaser also agrees to pay the Seller [10% of total estimated sale price] dollars ($_____) when the contract is signed as a refundable deposit to guarantee performance of Sections III and IV of this contract.

SECTION III

The Purchaser agrees to cut and remove said timber according to the following conditions:

1. Timber harvesting may begin on _[date]_ , and may continue until the termination date of _[date]_, unless an extension of time is requested and granted in writing. After this termination date all products remaining on the Seller's premises, cut or uncut, become the property of the Seller.
2. Trees cut for pulpwood shall be utilized to a minimum top diameter of at least 4 inches and those cut for sawlogs utilized to a minimum top diameter of at least 8 inches unless decay, branching, or stem deformity limits merchantability.
3. Stump heights shall be as low as practicable, but shall not exceed one-half their diameter.
4. Sawtimber and veneer logs shall be scaled by the Scribner Decimal C rule and pulpwood according to 128 cu. ft./cord for 8-foot wood and 133 1/3 cu. ft./cord for 100-inch wood.
5. Reasonable care shall be taken to protect the residual and neighboring stands from damage caused by logging activity.
6. Only timber designated in Section I shall be cut and removed. Whenever any undesignated trees are cut or needlessly damaged, the Purchaser shall pay for them at a rate of three times their scale value.
7. The Purchaser shall repair, at the Purchaser's expense, damage beyond ordinary wear and tear caused by Purchaser or Purchaser's agents to waterways, trails, roads, gates, fences, bridges, or other improvements on the Seller's property.
8. Locations of roads, landings, etc., shall be mutually agreed to by the Purchaser and the Seller or their agents.
9. Only nonmerchantable wood may be used for construction purposes in connection with the logging operation.
10. Best management practices for water quality shall be observed by the Purchaser for the duration of this contract.
11. The Purchaser shall remove all sale-generated debris within twenty (20) days of sale expiration, including machine parts, oil cans, paper, and other trash, and Purchaser's equipment and structures. Items not removed are deemed abandoned, become the property of the Seller, and may be removed or disposed of at the Purchaser's expense, including but not limited to the performance deposit.
12. Care shall be exercised at all times by the Purchaser and the Purchaser's agents against the start and spread of wildfire. The Purchaser agrees to pay for any and all damage and the cost of suppression of any fires caused by the Purchaser or Purchaser's agents.

SECTION IV

It is mutually understood and agreed by and between the parties hereto as follows:

1. The Purchaser agrees to save and hold harmless the Seller from any and all claims, penalties, or expenses of any nature, type, or description whatsoever, arising from the performance of this contract, whether asserted by an individual, organization, or governmental agency or subdivision. In furtherance of this clause, the Purchaser shall carry public liability insurance in the amount of $_________ and property damage insurance in the amount of $______________ . The Purchaser shall be responsible for the same insurance requirements on the part of any of its subcontractors.
2. Workers' compensation insurance, as necessary, and to at least the minimum extent required by law, shall be bought and maintained by the Purchaser to fully protect both Purchaser and Seller from any and all claims for injury or death arising from the performance of this contract.

3. This agreement shall not be assigned in whole or in part by either party without the written consent of the other party.
4. All timber included in this contract shall remain the property of the Seller until paid for in full.
5. The Seller guarantees property boundaries which are marked or otherwise designated. The Seller also guarantees that the Seller has full right and title to the timber included in this sale.
6. The Seller shall refund any performance deposit or notify the Purchaser of intent to retain said deposit within thirty (30) days of sale expiration. The Seller may suspend or cancel all operations for violation of any term of this contract by the Purchaser, and for cause may retain all monies deposited.
7. The Purchaser agrees that it is acting solely in the capacity of an independent party in carrying out the terms of this timber sale contract. It is agreed and acknowledged by the parties that the Purchaser is not an employee, partner, associate, agent, or joint venturer in any of the functions that it performs for the Seller. The Purchaser has a separate place of business.
8. The Purchaser agrees that it will furnish all materials, labor, equipment, tools, and other items necessary for the performance of this contract.
9. The Purchaser shall be responsible for filing its own legally required information returns and income tax forms.
10. The Purchaser has inspected the premises and knows and accepts it as being satisfactory to perform this contract without undue risk to person or property.
11. The Seller agrees that the Purchaser shall have sole control of the method, hours worked, time, and manner of any timber cutting to be performed hereunder. The Seller reserves the right only to inspect the job site for the sole purpose of ensuring that the cutting is progressing in compliance with the cutting practices established herein. The Seller takes no responsibility for supervision or direction of the performance of any of the harvesting to be performed by the undersigned Purchaser or of its employees or subcontractors. The Seller further agrees that it will exercise no control over the selection and dismissal of the Purchaser's employees.
12. In case of dispute over the terms of this contract, the final decision shall rest with an arbitration board of three persons, one to be selected by each party to this contract and a third to be selected by the other two members.
13. Special stipulations:

In witness whereof, the parties have set their hands on the dates shown below. We have read and understood all pages of this contract including any attachments, additions, or deletions.

Approved and agreed to by the Seller:______________________________

Signature of Seller Date

Approved and agreed to by the Purchaser: ______________________________

Signature of Purchaser Date

9

HARVESTING TIMBER

Most timber harvesting is done by professional loggers who have the equipment, knowledge, and experience necessary to conduct an effective and safe operation. However, you may want to harvest your own timber when you have only a small amount to cut, when you personally will use the timber, or when you expect little interest from commercial loggers in buying your timber. If you do choose to harvest your own timber, be sure to locate a market or personal use for the timber before harvesting. Recognize, too, that additional work and time spent in planning a timber harvest and conducting a safe operation will reward you with higher profits, a better quality residual stand, and protection of the environment.

Select trees to cut based your management objectives, stand conditions, and silvicultural principles described in Chapter 6. This chapter provides basic information on the harvesting process, including safety, building road systems, selecting harvesting equipment, using a chainsaw, and transportation.

SAFETY

Logging is one of the most hazardous occupations in the United States. While on the job, all workers, especially machinery operators, should wear a hard hat and safety toe shoes with anti-skid soles. Driver-operated equipment should have overhead canopies to protect the drivers from falling objects and rollovers and should carry fire extinguishers and first aid kits. No one should ride on equipment except in a belted seat under the canopy or in the cab. A productive and profitable operation can be realized only after safety has been given top priority.

BUILDING ROAD SYSTEMS

Few harvesting decisions have a longer impact than the design and construction of a forest road system. This system usually includes haul roads and skid trails. Haul roads usually are permanent features that provide wheeled vehicles access to specific points in the woodland for hauling logs or other management purposes. Skid trails generally are temporary, unimproved roadways that enable rubber-tired skidders or crawler tractors to transport logs from the interior of the woodland to landings where logs may be further processed and loaded onto trucks. A system of haul roads and skid trails must be carefully designed to minimize costs and environmental impacts (Figure 50).

Topography often will dictate the approximate layout of a road system necessary to bring timber out of the woods. Property lines, economic limits on skidding distances, and other features also influence the road system's design.

Haul Roads

Haul roads often are constructed by loggers, but landowners should understand the basic process and standards.

The slope of a road, called its grade, is expressed in percent. For example, a 10 percent grade is one that goes up or down 10 feet for every 100 feet of length. The road grade generally should not exceed 10 percent, although this limit may be exceeded for short distances up to a maximum of about 20 percent. Grades of 3 to 5 percent are the most desirable for effective drainage. Avoid long, steady grades that permit drainage water to flow down the road and cause erosion. On flat land avoid long, level road sections where water accumulates, causing road bed instability. To facilitate natural drainage, occasional changes to more level or steeper grades should be planned.

The potential for soil erosion and stream siltation is especially pronounced in areas with steep slopes and erodible soils. In these areas design the road to minimize its length.

On steep terrain, haul roads usually are placed on side hill positions (Figure 51). Such locations permit good cross drainage. They also provide the construction advantage of balanced cross-sections, which minimize earth moving.

Do not use streambeds for roads and avoid stream crossings whenever possible. Stream crossings can accelerate erosion and sedimentation and destroy plant and animal life in the stream.

Water draining from road surfaces usually carries sediment. Keep sediment out of streams and lakes by creating a filter strip, an area adjacent to the water body where soil disturbance is minimized so that vegetation and litter on the forest floor can trap sediment and protect surface waters from pollution. The width of the filter strip between a road and water varies according to particular site conditions. As a general guide, refer to Table 11.

Table 11. Filter strip width for various slopes.

AVERAGE SLOPE BETWEEN ROAD AND WATERCOURSE (PERCENT)	FILTER STRIP WIDTH (FEET)
0	25
5	35
10	45
15	55
20	65
25	75
30	85
35	95
40	105
45	115

SOURCE: Adapted from Simmons, F. C. 1979. *Handbook for Eastern Timber Harvesting (Stock No. 001-001-00443-0).* U.S. Government Printing Office, Washington, DC 20402. p. 141.

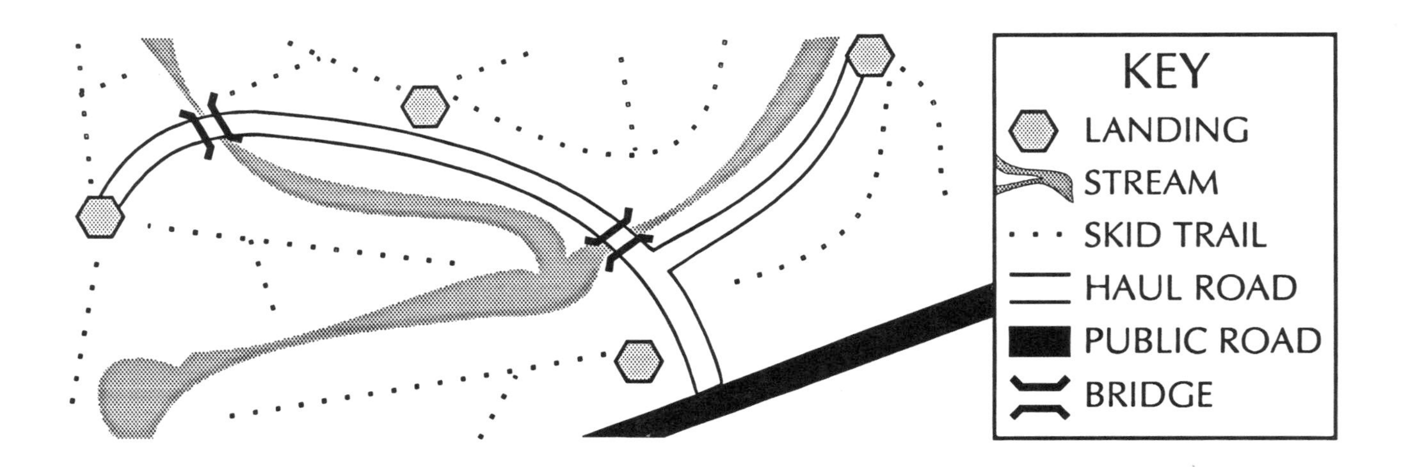

Fig. 50. Example of a plan for a system of haul roads, skid trails, and log landings.

Begin constructing a haul road by marking the proposed location, then clearing trees from the right-of-way. Move logs and limbs far enough off the right-of-way so that they will not interfere with construction or subsequent road use. Stumps that will be covered by a foot or more of fill may be cut low and need not be removed. Remove all other stumps and roots more than 3 inches in diameter.

Loose, exposed mineral soil around new roads erodes easily. Take care to place excess soil in stable locations and to compact any soil used as fill to minimize settling and reduce the entry of water. Siltation barriers such as straw bales and mulch help stabilize bare soils prior to establishment of vegetative cover.

Culverts should be used when roadways cross small drainages or when it is necessary to divert water from one road ditch to another. Install culverts large enough to handle the greatest expected runoff, or water may flow over the road, causing serious driving hazards and erosion. Keep culverts and ditches clear of debris to insure continuous drainage.

If the road will be kept open after logging, follow these simple rules to protect it: (1) keep travel to a minimum, (2) use the road only during dry weather or when the soil is frozen, and (3) inspect periodically and maintain as needed.

WATER QUALITY

! The site disturbance caused by timber harvesting is a potential source of nonpoint source (NPS) pollution (water pollution that originates from a dispersed source rather than from a pipe or other point source). The pollutants most often associated with timber harvesting are sediments, petroleum products, organic matter, and heat.

! A variety of techniques have been developed to protect water quality while still accomplishing important forest management objectives. These techniques collectively are called best management practices (BMPs). Every forest landowner and logger should become familiar with and use these practices. Many of them are described in this chapter, but for more complete information, refer to *Water Quality in Forest Management: Best Management Practices in Minnesota* listed in the reference section at the end of this chapter.

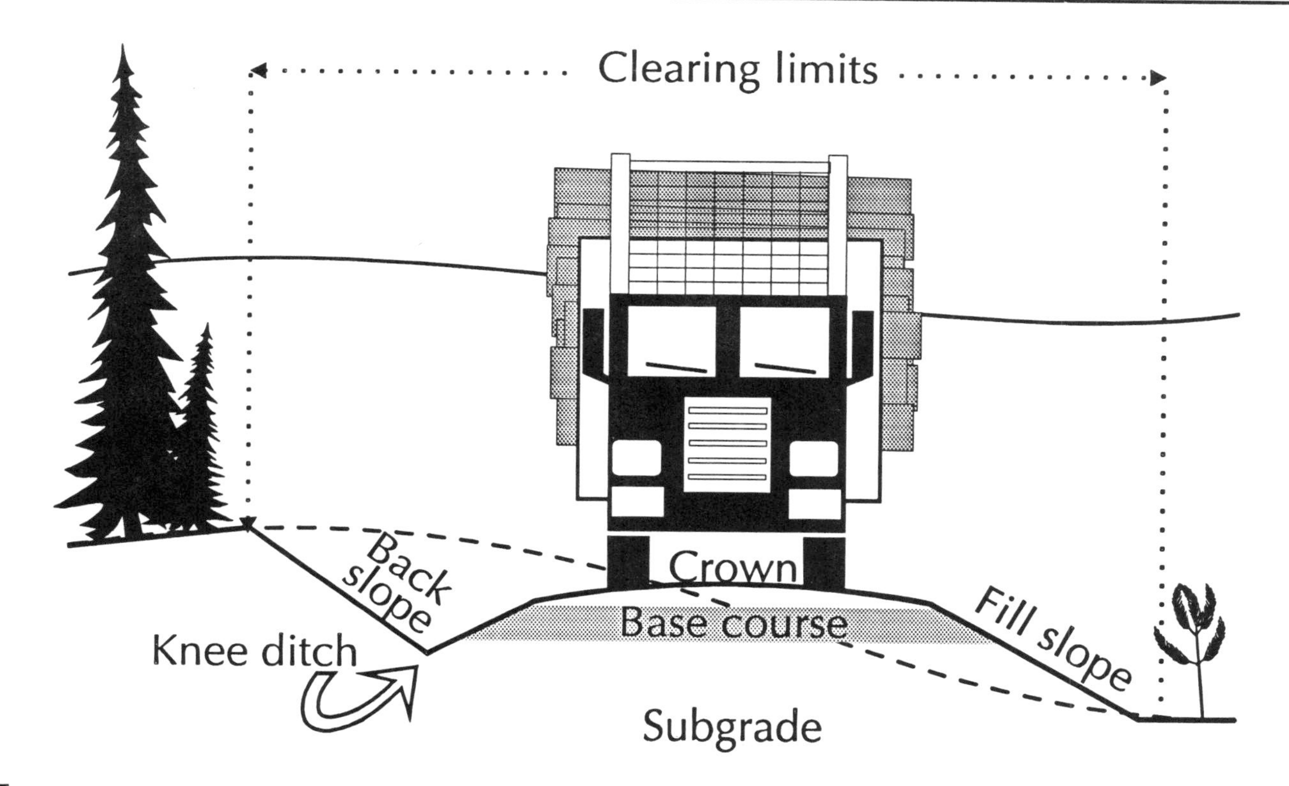

Fig. 51. Cross-section of a haul road on steep terrain.

Haul road maintenance is relatively minor with a properly located and well-constructed road system. Grade the road surface when necessary to fill in potholes and wheel ruts and reduce washboarding. Maintain a crown (raised center line) on the road to promote drainage into roadside ditches. Add gravel to soft spots. Remove roadside trees and brush so that sunlight can reach the road and help keep it dry and safe.

If roads will not be used for several years, replace culverts with water bars or other maintenance-free water control structures immediately after logging. *Water Quality in Forest Management: Best Management Practices in Minnesota* (see reference list at end of chapter) is a good source of information on closing inactive roads.

Revegetate roads, ditches, and landings with grasses and forbs to help prevent erosion. The seed mix and application rate will vary according to your climate and soil, but tall fescue or red fescue sown at a 10 to 15 pounds per acre rate generally provides adequate coverage and soil stabilization. The seed mixture can be altered to include food and cover plants that are beneficial to wildlife. Where possible, lightly disk and fertilize the bare roadbed before broadcast seeding.

Skid Trails

Skidding is the process of transporting logs from the point where the trees are felled to a landing, where they can be further processed or loaded onto trucks. Logs usually are dragged by a bulldozer or rubber-tired skidder, thus creating skid trails. Skid trails usually are not graded and need only a minimum amount of clearing. Depending upon management objectives and forest conditions, logs may be skidded over a fixed trail network or the logger could use a different skid route each trip to help knock down undesirable trees and shrubs, thus helping to clear the site for regeneration. Skidding generally should be restricted to distances of 1/4 mile or less. To protect soil and water, do not create any skid trails that run straight up and down hillsides, avoid streambeds and wet spots when possible (cross them at right angles when crossing is necessary), avoid rocky places, and make frequent breaks in grade.

Landings

Landings are areas used for processing and loading timber products onto trucks. Landings are busy places during a harvesting operation, producing big impacts on a relatively small area. Carefully select landing locations to provide for efficient timber removal and minimize adverse environmental impacts. Proper construction and maintenance of a landing is similar to that for skid trails.

Locate landings close to concentrations of timber. Choose locations with a slight slope so that water will drain away. Avoid steep slopes and low, wet areas where trucks cannot maneuver. Locate landings below ridge crests to reduce the need for steep, hazardous roads. The haul road approaching the landing should have a low grade. Landing size and shape will be influenced by the timber length, loading method, and type of hauling equipment.

If bucking, debarking, or other processing occurs on the landing, some limbs, bark, or other woody debris will remain at the completion of the logging job. Although the material could be piled and burned or left to decay, it also may be a convenient source of firewood.

Special care must be taken when storing petroleum products and maintaining equipment in woodlands. Designate a specified place for draining vehicle lubricants so they can be collected and stored until transported off-site for recycling, reuse, or disposal. Provide receptacles for solid wastes such as grease tubes and oil filters. Locate refueling areas away from water.

SELECTING HARVESTING EQUIPMENT

Many different timber harvesting systems are available. Each system includes a distinct mix of equipment that works together to harvest and remove timber while preparing the site for the next forest. Steps included in this process are felling the trees, removing limbs and tops (if done), transporting logs to a landing, cutting logs into shorter lengths (bucking), loading logs onto a truck or trailer, and transporting logs to another location, usually a mill. Increasingly, whole trees are being skidded from their point of growth to a landing for processing. This style of logging often results in very complete wood use and presents a clean site for regenerating the next forest.

Several important considerations influence the choice of timber harvesting equipment. Some of the most important are:

1. **Tree size** affects the size of equipment needed to handle the timber.
2. **Silvicultural prescriptions** (e.g., clearcutting, shelterwood harvesting, thinning) influence the choice of tree felling and skidding equipment.
3. **Topography**, especially ground slope, affects the type of equipment and method used to skid logs.
4. **Wood volume to be removed and time constraints** influence the preferred mix of equipment.
5. **Slash cleanup or noncommercial tree removal** may require specialized equipment.
6. **Fragile site** elements, such as unstable soil and proximity to water bodies and wetlands, may limit the size and type of equipment.

Harvesting equipment includes equipment used to fell trees and machinery needed to transport trees and products from the site. Specific types of equipment and their uses are described further in the sections that follow.

USING A CHAIN SAW

A chain saw is the most common tool used by woodland owners to fell, limb, and buck trees into usable lengths. Despite their wide availability and ease of operation, chain saws are linked to many accidents and severe injuries. Most chain saw injuries result from contact with a moving chain, while chain saw-related deaths usually occur when the operator is struck by trees or branches.

Reduce the potential for chain saw-related accidents by paying attention to three components of every accident—human, agent, and environment. The human component refers to your physical and mental condition while chain sawing. Although the agent that inflicts injuries usually is the chain saw, accidents also result from falling trees or limbs and from loss of balance, which can lead to dangerous falls. The environment in which you work, especially weather conditions, also can influence the likelihood of an accident. To prevent accidents, be physically fit, keep your mind focused on the task, be mindful of agents that could cause accidents, and be aware of the environment around you.

Safety Equipment

Before operating a chain saw, protect yourself with clothing and other gear designed to reduce the severity of any accident (Figure 52). The basic piece of personal protective gear is a hard hat. It should have attached earmuffs to reduce harmful noise and a screen-shield to protect your eyes from flying debris. Inexpensive ear plugs also are effective for protecting your ears from loud noise. Eyeglasses with safety glass also can substitute for a screen-shield, although lack of air circulation behind the lenses frequently causes fogging during exertion. Other protective gear should include sturdy work boots, leather gloves, and leg chaps. Be sure that chaps are constructed of material designed for protection from chain saws. Such materials are designed to slow or stop the chain if it strikes your leg, allowing you slightly more reaction time.

Fig. 52. Personal safety clothing and equipment for loggers.

General Operating Rules

Several general guidelines help make chain saw use more efficient and less dangerous. Learn to look up as you approach a tree before cutting. Overhead hazards can include utility wires, other trees, and dead and loose branches. Chain saws are designed to be run at full throttle. Always accelerate the engine before beginning a cut. Maintain your balance while operating the saw. Shift your position instead of overextending your reach and avoid reaching above shoulder height.

Kickback can result in severe injuries. It occurs when the chain strikes an object that stops its movement very suddenly and the engine power that was driving the chain thrusts the saw blade up toward the operator. Kickback occurs most often when trying to saw a log using only the saw tip or when saw teeth at the tip of the bar strike a hard object such as rock or metal.

To reduce the kickback hazard, saw wood with the long edge of the bar, not the tip. Keep your eyes on the saw tip (the "kickback zone") and make sure it does not contact hard objects. Grasp the saw firmly. If you are right-handed, grasp the saw's handlebar with your left hand and wrap your thumb around the handlebar. Keep your saw operating at peak efficiency. Sharpen and clean the saw often, and if it has a chain brake, test it regularly and adjust as needed. Your owner's manual will point out any special attention your saw needs.

Getting Started

Before you fuel the saw, clear the area around and under it of woody debris and other flammable material. Make sure the engine is cool, then add fuel and chain oil as needed. Move the chain saw at least 10 feet from the fueling area and be sure that it is firmly supported when you are pulling on the starter cord. Do not attempt to start the saw while you are standing and holding it unsupported because it could pivot and strike you.

Before making the first cut to fell a standing tree, consider various factors that can influence how, and even if, you should fell it. If tree diameter is more than twice the length of the saw bar, the tree requires special cutting techniques best left to specialists.

Before you first cut into a tree, determine the direction in which you want the tree to fall. A tree with a slight lean usually is easier to fell since the lean helps direct its fall. A tree with severe lean can be dangerous to fell because the tree will begin to fall in the leaning direction before it has been completely severed at the stump. It will then split at the base, causing what is known as a "barber chair." Barber chairs can fly upwards, striking the operator with terrific force.

The soundness of a tree can affect its direction of fall. Evaluate soundness by looking for signs of decay including loose bark, fine "sawdust" particles at the base of the tree, or large holes in the trunk at any height. Trees usually decay from the center out, sometimes leaving only a shell of sound wood.

Distribution of the tree's crown also can affect felling direction. Check for large limbs, snow or ice accumulations, or uneven distribution of the crown.

Consider weather conditions, especially wind speed and direction, before choosing a felling direction. The effect of wind is more pronounced on trees with large crowns. Avoid felling on very windy days.

Evaluate the general terrain in the working area to gauge its effect on the felling direction and to determine a safe escape route to use as the tree begins to fall.

Look for nearby coworkers, vehicles, powerlines, and other objects that should be avoided when felling the tree.

The last step before beginning to cut is to clear brush from around the tree's base. This allows greater freedom of movement while sawing and makes it easier to step away once the tree begins to fall.

Directional Felling

During a timber harvest, the usual practice is to fell trees in a herringbone pattern adjacent to the skid trail (Figure 53). This practice reduces skidding time, timber breakage, and damage to residual trees. It also makes limbing and bucking easier.

Chain saw felling consists of two basic cuts, the undercut and the backcut (Figure 54). The undercut removes a wedge-shaped piece of trunk from the side toward which the tree will fall. Only one-fourth to one-third of the tree's diameter

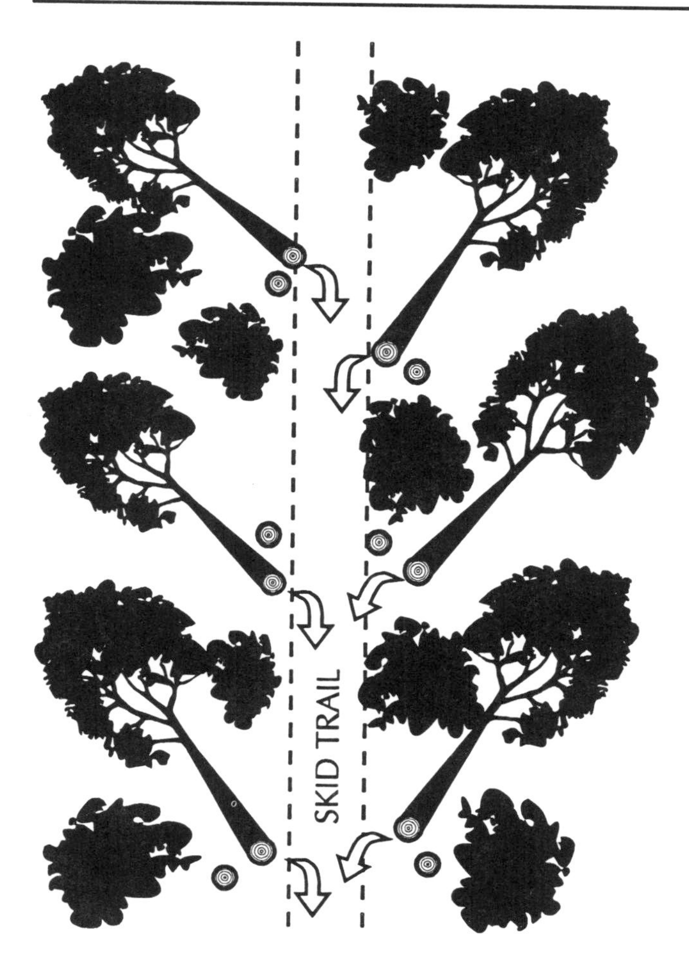

Fig. 53. Fell trees in a herringbone pattern along skid trails.

should be removed with the undercut. The back-cut, made on the opposite side, lets the tree fall.

The conventional undercut is made by first sawing the lower horizontal face and then sawing the upper face at a downward angle to meet the first cut. A newer style of undercut uses angled top and bottom faces to form a 90-degree angle. Use the new style of undercut whenever possible because the faces of the cut do not close until the tree is on the ground, giving a longer period of control over the tree's fall. Be sure that the cuts meet precisely. If one cut travels into the trunk too far, cut the other just deep enough to meet it.

The location of the undercut determines stump height. Trees should be cut as low to the ground as practicable to harvest all of the usable wood. Softwood species often can be cut within 6 inches of the ground, but hardwoods may need to be cut higher to avoid the basal flare of roots.

The horizontal backcut is made from the opposite side of the tree, about 2 inches above the V of the undercut. Be sure to stop before reaching the undercut. About 2 inches of uncut wood, called the "hinge," should remain to help guide the direction of fall. The hinge also helps keep the butt of the tree attached to the stump, reducing the chance of its striking you.

If you need to pull a tree slightly away from its natural lean, leave a hinge of so-called "holding wood" that is wider on the side toward which you

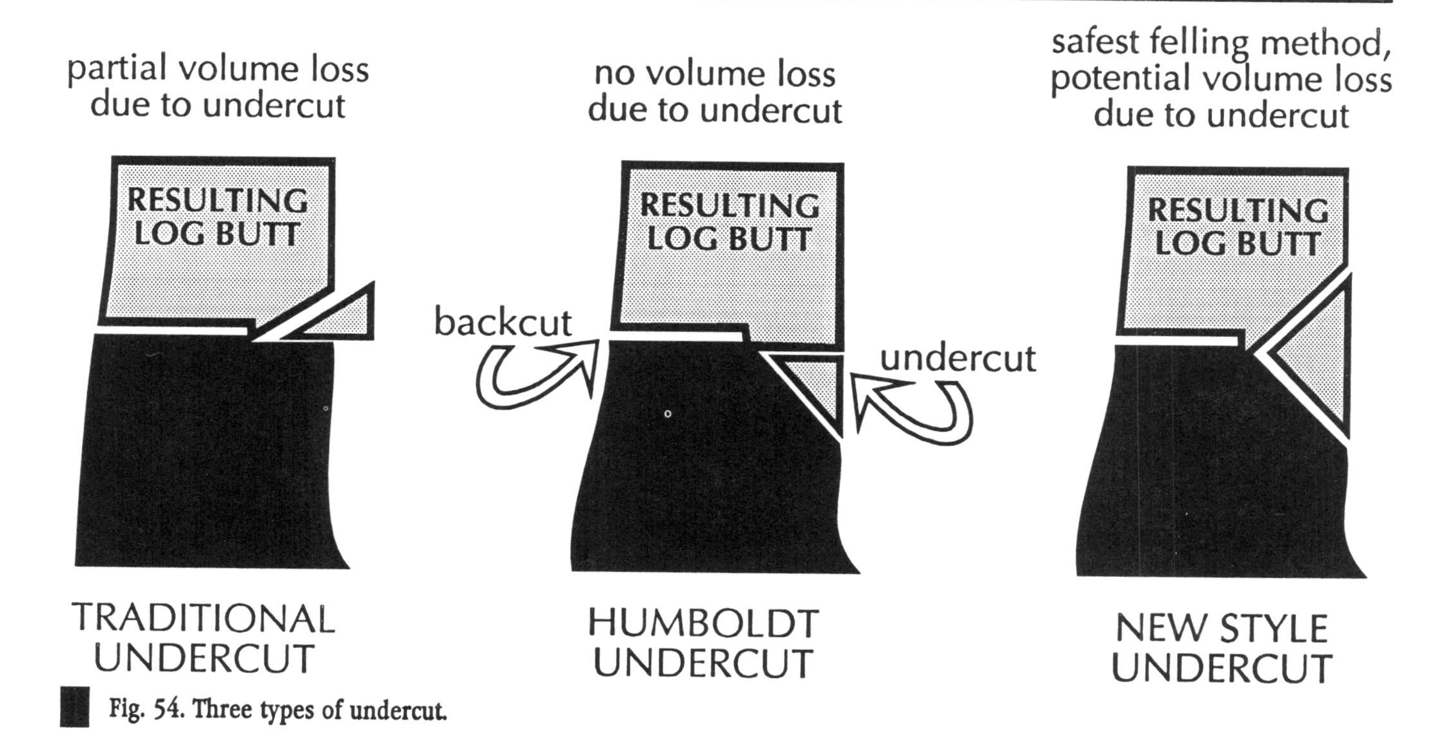

Fig. 54. Three types of undercut.

want the tree to fall (Figure 55). Larger trees sometimes can be forced to fall in a particular direction by driving wooden or plastic wedges into the backcut behind the saw (Figure 56).

When the tree begins to fall, retreat in a diagonal direction away from the direction of fall (Figure 57). Use a predetermined escape route and keep your eye on the falling tree for any developing problems.

If you are not very experienced, start on smaller trees to practice the basic felling cuts. Avoid felling a tree if it is in a risky situation. Secure an experienced helper or hire a professional for dangerous work. Never work alone. Avoid cutting trees when you are physically or mentally fatigued.

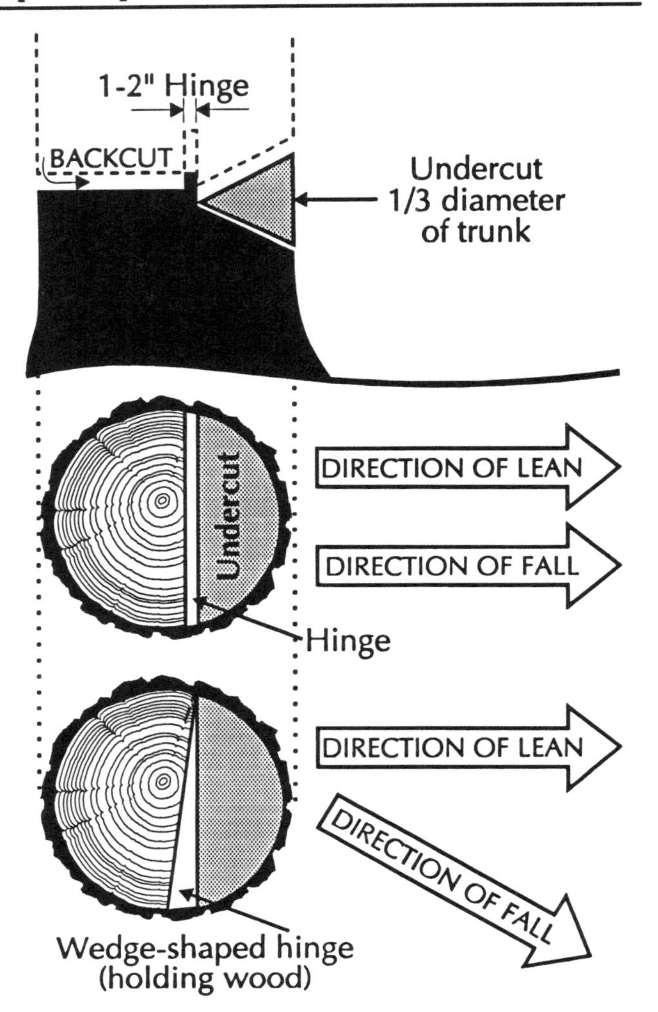

Fig. 55. Leave a hinge of wood to control the direction of fall.

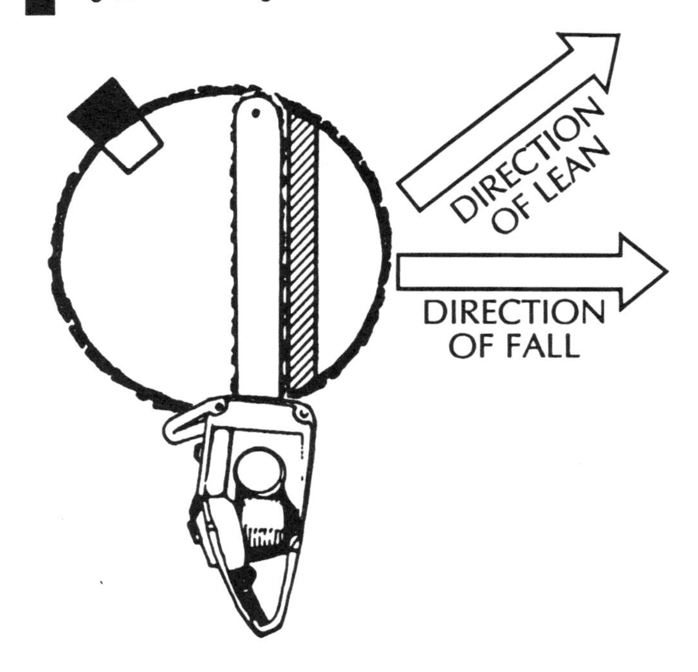

Fig. 56. A wedge driven into the backcut can force a tree to fall in a particular direction.

Removing Limbs and Bucking

Commercial loggers sometimes drag whole trees from the woodland, but landowners who cut their own trees usually remove the limbs and buck the tree into smaller pieces before dragging logs away. The length of these pieces depends on product specifications and the location of log defects. Leave a 3-to 6-inch trim allowance on sawtimber and veneer logs.

You can buck a fallen tree that lies flat on the ground into shorter logs simply by cutting from the top of the log down to the ground. Be careful not to cut into the ground or you could dull your saw and create dangerous flying debris.

Often fallen trees are supported at both ends by the ground and the tree's own limbs. Under these conditions a trunk experiences forces of tension on the underside and compression on the top. Tension wood is being stretched. Cutting into tension wood will not bind the saw. However, compression wood is being squeezed; if the saw cuts too far into the compression area, the gap will close and bind the saw.

To cut a log supported at both ends, first cut part way through the compression wood on top of the log. Then cut through the tension wood on the bottom of the log until the two cuts meet (Figure 58).

If a log is supported only at one end, compression wood occurs on the lower side. In this situation, first cut part way through the log from the

Fig. 57. When the tree begins to fall, step away in a diagonal direction from the falling tree.

bottom. Make the final cut from above, allowing the pieces to separate (Figure 59).

To remove large limbs you will also need to identify tension and compression sides. This is especially important for limbs on the underside of a fallen tree, which may be under severe stress.

When working on steep hillsides, always work from the uphill side of a fallen tree because logs can easily roll after bucking.

TRANSPORTATION

After trees have been felled, they need to be transported to the landing, loaded on a truck or trailer, and hauled to a mill or other site for processing.

Skidding and Forwarding

Short logs, tree-length logs, or whole trees may be transported from where they were cut to the landing either by skidding (dragging them on the ground) or forwarding (carrying them completely off the ground).

Logs can be transported to the landing using a four-wheel-drive rubber-tired skidder, steel-tracked crawler tractor, forwarder, or farm tractor (Figure 60). A rubber-tired skidder is lighter weight, less expensive, and faster, and provides better traction over rocks than a crawler tractor. It also can travel over paved highways. The crawler tractor, on the other hand, exerts less pressure per square inch on the ground surface, provides better traction in mud and slippery soils, and is less likely to damage the site through soil compaction or deep ruts. It also can be used in road construction and other earth-moving tasks.

Farm tractors sometimes are used for skidding, but they usually need modifications to become effective, safe skidders. A winch connected to the tractor's power take-off is an important asset. Some also require shields beneath their bodies to protect them from damage by high stumps and rocks.

Forwarding machines have a bunk for holding a load of logs and often are equipped with a hydraulic loading boom.

Loading

Lifting logs from the ground to a truck bed and securing them for safe transport is a difficult and dangerous task. Many loading systems have been devised. In hilly country skidways of logs once were built out from hillsides to store logs above ground level. A truck would pull up next to the skidway and the logs would be rolled onto the truck. In other cases the cross-haul method was used. It involved pulling the logs up a pair of planks by means of a cable sling attached to a power source on the opposite side of the truck bed. Most loading today, however, is done with hydraulically controlled knuckle-boom loaders mounted either on the hauling truck or on a separate vehicle. A front-end loader is another option for loading, but it more often is found in mill yards than in the woods.

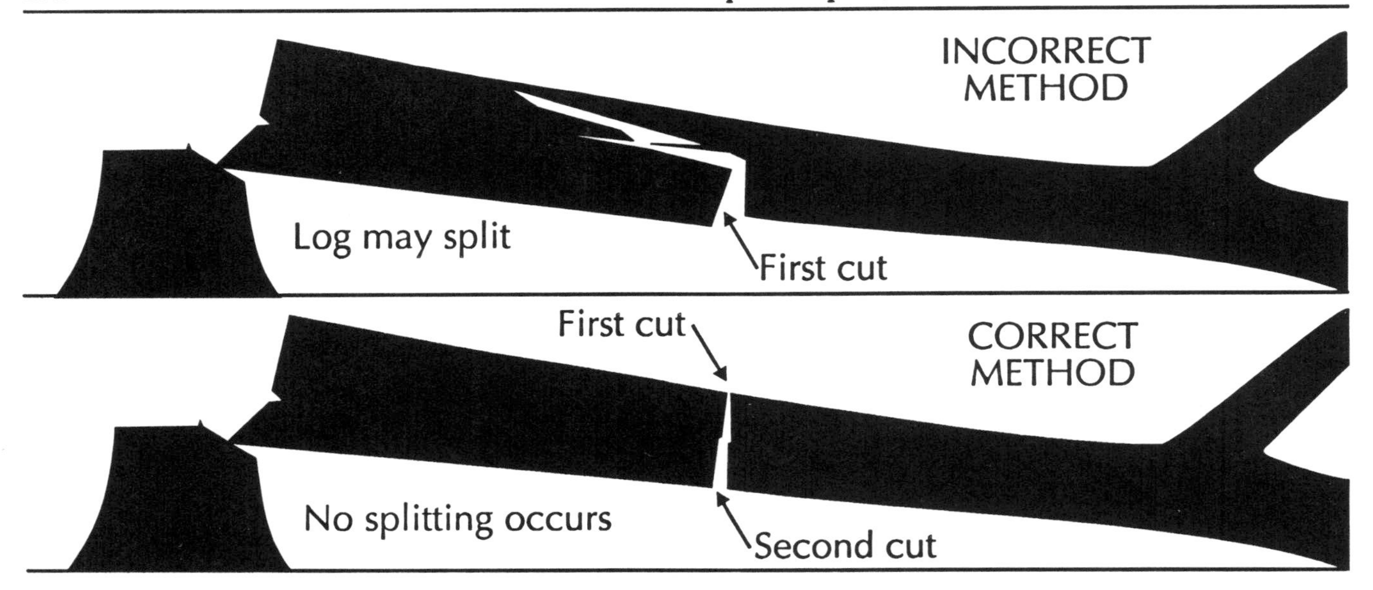

Fig. 58. Bucking a log that is supported at both ends.

Hauling Logs

Practically all logs transported from the logging operation to the processing mill are hauled by trucks. Some of this hauling may be done by an independent contractor and some may be done by the receiving mill. You also can haul your own logs to the mill, if you have suitable equipment and skills. All logs should be fastened securely to the truck and should fit within width, length, and weight requirements set by state laws.

Suggested References

Garland, J. J. 1983. *Designated Skid Trails Minimize Soil Compaction (Extension Circular 1110).* Oregon State University, Corvallis, OR 97331. 6 pp.

Garland, J. J. 1983. *Felling and Bucking Techniques for Woodland Owners (Extension Circular 1124).* Oregon State University, Corvallis, OR 97331. 12 pp.

Pope, P. E., B. C. Fischer, and D. L. Cassens. 1980. *Timber Harvesting Practices for Private Woodlands (FNR-101).* Purdue University, Cooperative Extension Service, West Lafayette, IN 47907. 8 pp.

Reed, A. S., R. A. Aherin, and L. Schultz. 1986. *Safe Chainsaw Operation (NR-FO-2487).* University of Minnesota, Minnesota Extension Service, 3 Coffey Hall, St. Paul, MN 55108. 6 pp.

Simmons, F. C. 1979. *Handbook for Eastern Timber Harvesting (Stock No. 001-001-00443-0).* U.S. Government Printing Office, Washington, DC 20402. 180 pp.

Minnesota Department of Natural Resources, Division of Forestry. 1990. *Water Quality in Forest Management: Best Management Practices in Minnesota.* 500 Lafayette Road, St. Paul, MN 55155. 104 pp.

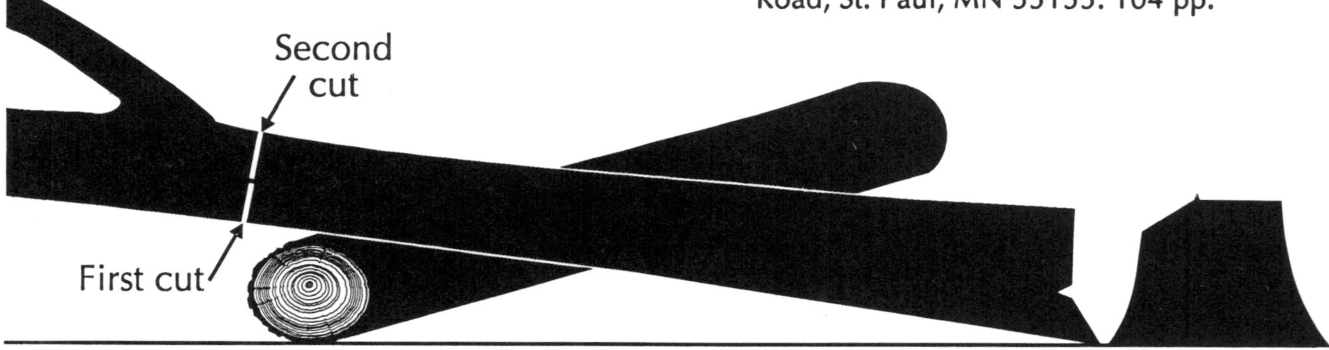

Fig. 59. Bucking a log that is supported only at one end.

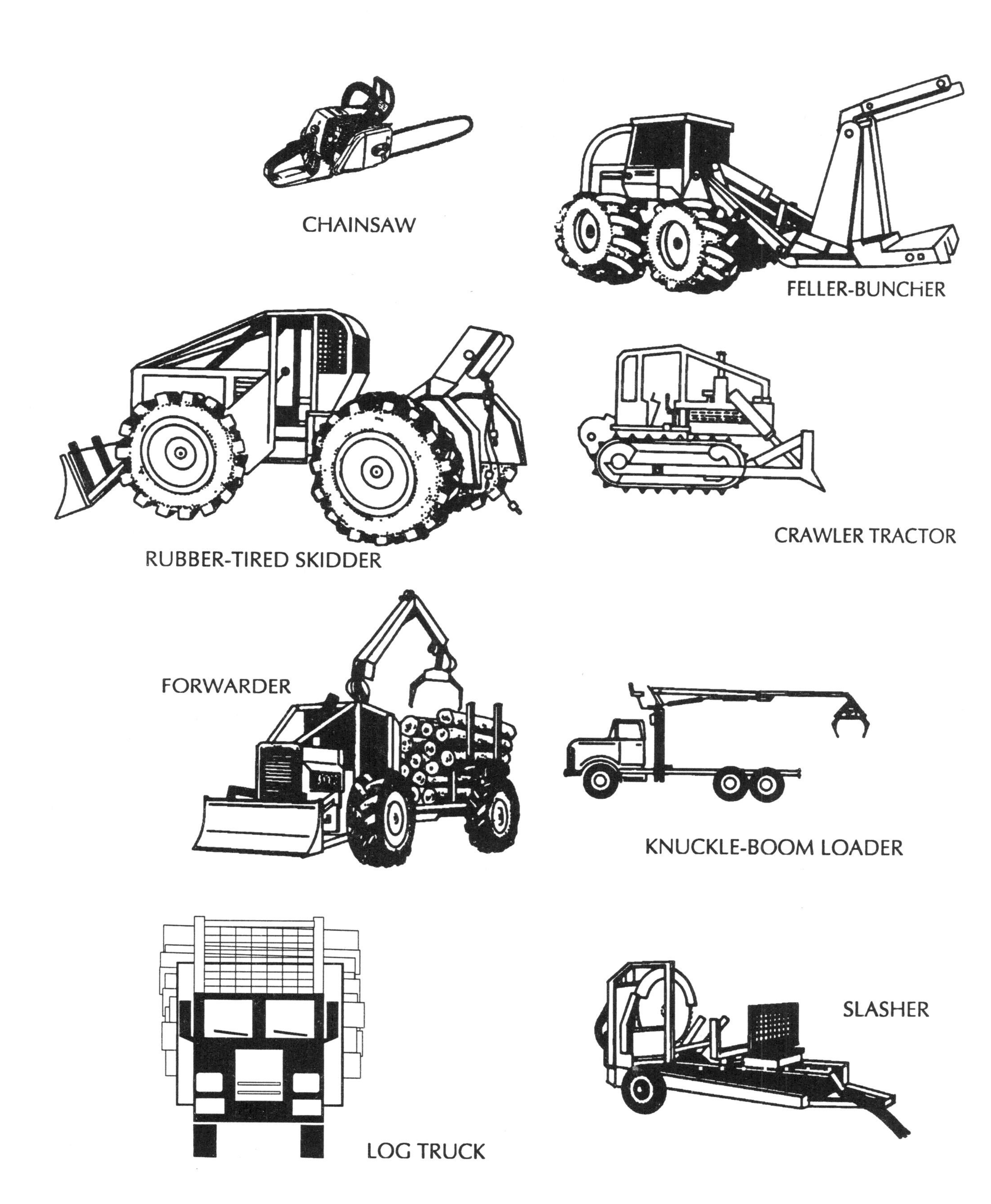

Fig. 60. Logging equipment.

10

MANAGING A WOODLAND FOR WILDLIFE

Woodlands provide habitat for a wide variety of wildlife species. Some are year-round residents; others use our northern woodlands during the warm seasons, but migrate south in winter. Each species needs living space as well as a distinct type and arrangement of shelter, food, and water. Although some woodland owners believe that wildlife need woodlands that are protected from harvesting, fire, insects, and diseases, that is true only for a few species. Through active woodland management you can improve the habitat for a wide variety of wildlife species, or favor particular species.

SITE QUALITY

The quality of your habitat is limited partly by overall site quality. Site quality affects both the diversity of wildlife species and the numbers of animals. Some of the factors influencing site quality, as discussed in Chapter 3, are soil depth, texture, moisture, and fertility; topography, including slope and aspect; and climatic factors such as length of the frost-free growing season, cold temperature extremes, precipitation amount, and duration of droughts.

A site that is poor for growing trees, also is poor for supporting diverse wildlife species and large numbers of wildlife. Good growing sites for trees also are good sites for wildlife because they provide the vegetation needed for food and cover.

SPACE

Each animal species needs a certain amount of space (home range) where it can find the food, water, and cover needed for survival. Within an animal's home range, there may be a core area where it spends most of its time. The location of core areas may change seasonally. The size of an animal's range depends partly on habitat quality; the higher the quality, the smaller the range needed to support the animal. Depending on the size of your woodland, you may be able to provide all of the habitat for wildlife species with small ranges, but only a portion of the habitat for more wide-ranging species. To attract wildlife that roam over large areas, consider the habitat on surrounding properties and then provide the most limiting component of their required habitat.

FOOD AND SHELTER

Wildlife need food and shelter from predators and the weather. Each species has its own food and shelter requirements.

Tree cavities provide homes for squirrels, raccoons, and several bird species. While such den trees often have decay, making them unsuitable for timber, live den trees with small crowns do not take up much growing space and are very useful to wildlife.

You can encourage cavity formation in trees by selecting a limb at least 3 inches in diameter and pruning it off about 6 inches from the trunk. Over time the limb will decay and a cavity may form there. Elm, ash, and basswood are especially prone to form natural cavities. You also can bore a hole at least 2 inches in diameter to the center of a living tree. Drill the hole just under a limb 3 inches or larger in diameter. The hole will enlarge as the wood decays and provide a cavity. Do not bore holes in oak trees from mid-April through mid-July or the oak wilt fungus could become established and kill the tree.

Dead standing trees, called snags, may be hollowed out by woodpeckers for nest sites. Snags also harbor insects beneath their bark and in their soft, dead wood that can be captured by birds. Snags do not interfere with growth of surrounding trees and should be left in hardwood stands for wildlife. Do not leave snags in conifer stands, however, since freshly killed conifers may serve as breeding sites for bark beetles that can infest and kill nearby trees.

Snags can be left in clearcuts or created during TSI operations that involve thinning, culling, and weeding. In hardwood stands you can create snags by girdling the stem of a tree or by a combination of girdling and herbicide application (refer to Chapter 5).

Snags are useful to the largest variety of wildlife when they are at least 12 inches DBH, although some bird species can use holes in trees as small as 4 inches DBH. Trees with relatively hard wood such as oak, maple, basswood, ash, and elm are most useful to wildlife. Aspen snags are good if trees are at least 12 inches DBH.

Both hard snags and soft snags are useful to wildlife. Hard snags are recently killed trees that still retain some of their branches. Soft snags are dead trees that have lost their branches and have wood that is turning soft from decay. By killing a few trees over 12 inches DBH each year, you can replenish the supply of hard snags.

Over time you can try to achieve the following optimum number of snags for a 20-acre woodlot: four to five snags over 18 inches DBH, thirty to forty snags over 14 inches DBH, and fifty to sixty snags over 6 inches DBH.

When snags fall over, leave the logs to decay. Rotting logs recycle nutrients into the soil and continue to provide food and cover for birds, small mammals, reptiles, and amphibians.

In a woodland some animals primarily inhabit the overstory, and others favor the understory. You can diversify habitat by developing stands that have layers of vegetation from the ground up to the highest canopy. If you have a high overhead canopy that blocks the sun and inhibits understory development, thin the stand lightly (leaving 60 to 80 percent crown cover) to stimulate growth of understory plants.

Some animals that live on or near the ground find protection in thickets created by shrubs or dense stands of young trees. Thickets may occur naturally along woodland edges where the additional sunlight encourages plant growth. Thickets can be encouraged by thinning the overhead canopy to permit sunlight to stimulate plant growth or by harvesting under group selection or clearcut systems. Root-sprouting tree species such as aspen form dense thickets following clearcutting.

Dense conifer stands shelter deer in the winter, and the dense foliage provides cover for certain bird species. Cones hold seeds eaten by squirrels and birds.

Brush piles provide cover and nesting sites for various mammals. They are most effective when placed along the edges of woodlands and fields and in clearcuts with second-growth vegetation. Brush piles should be at least 5 feet high and 12 to 15 feet in diameter. They will last longer if they have a base of large logs, stumps, or rocks with smaller brush piled on top. Sections of culverts, field tiles, or hollow logs placed at the base will provide dens.

Brush piles also are useful to reptiles and amphibians if located on the edge of a pond or lake with part of the brush submerged.

A mixture of stands of different ages and densities will appeal to some wildlife. Ruffed grouse, for example, prefer aspen stands 4 to 15 years old for brood cover, 6- to 25-year-old stands for fall and spring cover, and older stands for food and nesting cover. To attract grouse, keep stands small (2 to 10 acres) and create a mix of sapling-, poletimber-, and sawtimber-sized stands interspersed with permanent openings.

Some wildlife species find the shelter and food they need in large, contiguous areas of old-growth woodland. Old growth woodlands commonly are defined as having an average stand age at least one and one-half times the normal rotation age. Such areas are becoming scarce in the Upper Midwest. Consider leaving old-growth woodlands uncut, especially if they have poor road access, are isolated by water, have low economic value, occur in eagle and osprey nest buffer zones, or occur where aesthetics are important.

When clearcutting a large area, make the cutting block narrow and irregular in shape because few wildlife species will use the center of a large clearcut. If you plan a wide clearcut, leave uncut corridors of trees across the clearcut and along water courses to serve as travel lanes for wildlife.

Thin hardwood stands, especially those with oaks and other mast-producing species, to encourage crown expansion and subsequently increase mast production.

Permanent forest openings of 1/2 to 10 acres can be valuable to many wildlife species. Openings could be old fields, log landings, or gravel pits. Such openings may offer forage for deer, insects for birds, and habitat for small mammals preyed upon by raptors and foxes. Bare, sandy soil offers dusting sites for birds that need dust baths to control external parasites.

Following a harvest, reseed forest roads and landings to provide green forage. Seed mixtures vary depending on the site. Contact a forester, wildlife manager, or soil conservationist for recommendations.

WATER

Most wildlife species need surface water for drinking. Ponds and marshes also are the principal habitat for many species of ducks, frogs, turtles, snakes, and salamanders. If there is no permanent surface water on your property, consider adding a pond. Planning assistance is available from your local Soil and Water Conservation District or U. S. Department of Agriculture, Soil Conservation Service office.

MANAGING FOR SELECTED SPECIES

Activities described above will enhance your woodland for a wide range of wildlife species. You may, however, wish to improve habitat for particular species. The following sections describe management that you can perform to encourage selected species to make greater use of your property.

White-tailed Deer

White-tailed deer range over several square miles. To manage for white-tailed deer, consider the habitat provided on surrounding properties and attempt to provide the part of the habitat that is in shortest supply.

White-tailed deer need a combination of escape cover, winter shelter, and food. Escape cover

Fig. 61. White-tailed deer.

is provided by sapling-sized stands and brushy areas. Conifer stands with a dense canopy make excellent winter shelter. Woody browse may be provided in seedling stands that develop following clearcuts or in mature stands with a woody understory. Herbaceous vegetation used for food can be found in permanent openings or beneath timber stands such as aspen with moderate crown cover. Preferred browse plants vary locally, but may include dogwoods, mountain and red maple, ash, northern white-cedar, eastern hemlock, filbert (hazel), birches, willows, mountain-ash, sumac, and aspen. Acorns also are a prime food. Deer will eat a variety of agricultural crops, but such crops are not essential for their diet.

Here are some habitat management suggestions for deer:

- Dense escape cover is best created by clearcutting or shelterwood harvests that result in dense sapling stands. Clearcuts should be less than 400 feet wide because deer will not feed in the center of larger clearcuts.
- To provide winter cover in the Upper Midwest where heavy snowfall limits deer mobility and food supplies, plant or maintain conifer stands with a dense canopy.
- Manage older hardwood stands to increase mast production. Thin hardwood stands to permit mast-producing trees to continue crown expansion. During reproduction harvests and TSI operations, favor mast-producing species, especially oaks.
- Encourage reproduction and development of aspen stands. Young aspen clearcuts provide browse, sapling stands offer escape cover, and older stands usually have a dense understory of shrubs for browsing.
- Maintain 10 percent of the forest area in permanent openings to provide herbaceous vegetation for food.

Ruffed Grouse

A breeding pair of ruffed grouse can survive in about 10 acres of woodland if all their habitat requirements are met. Grouse need dense sapling- to pole-sized stands for cover, but feed mostly in older stands. They eat flower buds, especially those of aspen, birch, and hazel, in winter; aspen catkins, green herbaceous vegetation, and insects in the spring; and insects, green herbaceous plants, fruits, and berries in summer and fall.

Manage for grouse by creating a mix of age classes of timber in close proximity. Divide the woodland into cutting blocks of 2-1/2 to 10 acres and harvest these blocks by clearcutting. Each 40 acres should have at least four cutting blocks ranging in size through seedling, sapling, poletimber, and small sawtimber classes.

Aspen stands can be managed easily for grouse because when clearcut they produce dense stands of root suckers that grouse use for cover.

Mixed hardwood stands are suitable for grouse if dense stands can be reproduced. Increase the aspen component when possible.

While grouse will use conifer stands for winter cover, large conifers (especially pines) provide roosts for hawks and owls, which prey on grouse.

Fig. 62. Ruffed grouse.

Grouse can survive the winter well without conifers if there is adequate snow cover for snow burrows.

Following harvests remove as much debris as possible, because logs on the ground offer cover for ground predators such as foxes. Leave a few large logs where they will be surrounded by a high density of vertical stems so that grouse can use them for drumming sites.

Fig. 63. Woodcock.

Woodcock

Woodcock are migratory but spend spring, summer, and early fall in the North. They feed mainly on earthworms, which they find with their long bills in soft, moist soil along streams or in other wet sites. Alder and aspen stands on moist sites are good feeding areas. Their leaf litter creates an environment favored by the earthworms that woodcock prefer. Like grouse, woodcock prefer dense sapling- to pole-size stands for daytime cover, but they may roost at night in large open fields or clearcuts.

Woodland management practices recommended for ruffed grouse are likely to encourage woodcock. However, woodcock also need "singing grounds" during spring courtship—spots of bare ground, 15 to 20 feet in diameter, surrounded by shrubs or trees no more than 10 feet tall for a radius of 10 to 30 yards. Old fields, log landings, and forest roads may provide singing grounds. One singing ground per 10 acres is sufficient.

Gray and Fox Squirrels

Gray and fox squirrels may find all their habitat needs in 2 acres. They seek concealment in dens in hollow trees or in leaf nests they make high in tree crowns. Squirrels feed primarily on mast, tree buds, bark, fruits, and berries. Their primary habitat is sawtimber-sized hardwood stands.

To favor squirrels, manage hardwood stands on long rotations, retain den trees and vines, encourage mast production by thinning around the crowns of mast-producing trees, and favor mast-producing tree species when regenerating the stand. Harvest stands by group selection in order to retain continuous forest cover while creating openings large enough to encourage reproduction of oaks and other mast-producing trees that are intermediate in shade tolerance. You could plant oaks and walnuts on suitable sites in old fields or large forest openings.

Fig. 64. Gray squirrel.

Wild Turkeys

Wild turkeys are common in large oak-hickory woodlands in southern parts of the Lake States and in states to the south. Flocks range over about 4 square miles. Turkeys need open hardwood forests interspersed with permanent openings for food and cover. They roost in large trees at night, but during daylight forage on the ground for mast, fleshy fruits, green leaves, and insects.

To encourage wild turkeys, follow the recommendations in Chapter 6 for managing oak-hickory woodlands. Maintain part of the forest in open stands of large mast-producing species, especially oaks. Encourage acorn production by thinning around the crowns of oak trees. Encourage oak reproduction. Reproduce stands by small clearcuts or group selection. Keep livestock out of the woodland. If possible, leave food plots of grain near woodlands during winter to supplement the natural food supply.

Fig. 66. Wild turkey.

Songbirds

Songbird habitat varies by species. Depending on the species, songbirds may feed on the ground, in the understory, on tree trunks, or high in the canopy. Nesting sites are similarly variable.

Fig. 65. Winter wren.

Attract a diversity of songbirds by offering a diversity of habitats. Some species are attracted to woodland edges, but others require large stands of contiguous forest.

You can attact edge-oriented species by intermingling stands of different tree species, ages, and densities. But, keep in mind that large areas of contiguous forest are becoming scarce and the migratory birds that need them are declining in numbers worldwide. There are a lot of forest edges in the Midwest; more unbroken forests are needed. Create snags and retain den trees. Develop permanent forest openings. Provide both conifers and hardwoods.

Suggested References

Gullion, G. W. 1984. *Managing Northern Forests for Wildlife (Misc. Journ. Series No. 13,442).* Minnesota Agricultural Experiment Station, University of Minnesota, St. Paul, MN 55108. 72 pp.

Henderson, C. L. 1987. *Landscaping for Wildlife.* Minnesota Department of Natural Resources, 500 Lafayette Road, St. Paul, MN 55155. 145 pp.

11

AESTHETIC CONSIDERATIONS

Many woodland management practices affect the appearance of a woodland. Whether the resulting appearance is pleasing or displeasing depends on who does the observing. For example, some people prefer the random spacing and mixture of tree species in stands that developed naturally, while others prefer well-manicured, single-species stands planted in straight rows. There are many other ways in which our opinions about aesthetics differ.

You should plan management practices with an eye toward aesthetics because the public at large cares about the appearance of your woodland at least as much as you do. Neighbors, visitors, and tourists may not recognize the biological or economic factors that influence your decisions to harvest or regenerate trees, but they will judge your land stewardship by the physical appearance of your woodland. The future of some management practices may be influenced, possibly even regulated, as a consequence of public pressure to maintain attractive woodlands.

Following are some management practices that will affect the appearance of your woodland. Choose those that match your visual preferences.

LANDSCAPE MANAGEMENT

Portions of your woodland that are visible from public roads or waterways require special attention. Try to maintain a natural appearing landscape in those areas. Begin by considering the part that your land plays in the overall landscape of the area.

In areas where there are large expanses of unbroken forest, trees may become monotonous to travelers. There you can provide a visual break and a more interesting landscape by creating a mixture of stands with different tree sizes and species. In such areas permanent openings also could be created to permit wildlife viewing and to attract a greater diversity of wildlife.

You can create or enhance a scenic vista by felling trees in the foreground or by thinning the stand or pruning lower limbs to permit a view of the broader landscape beyond. On a smaller scale you can clear sight lines through heavy undergrowth to draw attention to picturesque trees, rock formations, streams, lakes, or other scenic attractions.

Manage the tree species composition of your woodland to encourage trees with special visual appeal because of their trunk shape, blossoms, bark color, fall foliage color, or other characteristics. For example, white birch and aspen have white bark that is especially attractive when contrasted with green foliage on spruce and fir trees. In the fall, sugar maple leaves may turn brilliant red while aspen leaves turn golden. Just as gardeners plan flower gardens to bloom all summer, you can manage your woodland to create visual appeal throughout the year.

Some people prefer woodlands with a high canopy but little understory vegetation. You can encourage these open, parklike conditions by maintaining a dense overstory that shades the ground and discourages understory vegetation. Dense conifer stands (e.g., pine, spruce, fir) block sunlight more effectively than deciduous trees. Deciduous trees such as sugar maple and oak have dense crowns that block sunlight more effectively than aspen and birch.

To block vision into a woodland and screen an objectionable view or create privacy, encourage shrubs and ground vegetation by lightly thinning a dense canopy. Try leaving 60 to 80 percent crown cover. For better year-round screening, establish spruce, fir, or pine along the woodland edge.

TIMBER HARVESTING

In most areas only 1 to 3 percent of the whole forest is subject to harvesting in a single year, so the impact of timber harvesting on a large landscape is relatively minor until the effects accumulate over a long period of time. However, in the small areas where timber harvesting does occur, it can dramatically alter the appearance of a woodland. To minimize the visual impact of harvesting, plan harvest and regeneration systems that quickly regenerate stands. Each harvest and regeneration system described in Chapter 4 has a different visual affect.

Harvest stands by the single-tree or group selection method to produce an uneven-aged woodland and avoid the temporary barren appearance of clearcuts. Keep in mind, however, that the selection method regenerates mainly shade-tolerant tree species.

The difference between group selection and clearcutting is a matter of scale. Clearcuts often are described as larger than 2 acres whereas group selection cuts are smaller.

You can soften the visual affect of clearcuts by making them as small as possible; by cutting them in narrow, irregular shapes; by feathering (thinning into) the borders where they adjoin stands of older trees (Figure 67); by shaping them so only a small portion is visible from any one viewing point; by leaving tree islands in the clearcuts or corridors of uncut trees across clearcuts; by shaping them to follow major land contours rather than cutting across the landscape (Figure 68); and by leaving screens of uncut trees between the clearcut and public roads, trails, or waterways.

Some people prefer the appearance of complete clearcuts where no trees are left standing, whereas others prefer clearcuts that retain scattered live trees. Scattered live trees break up the monotony of large clearcuts and provide vertical habitat for birds. Some songbird species benefit from the residual trees; however, those trees also serve as raptor perches, giving avian predators an advantage. If rodents are likely to cause regeneration problems by eating tree seeds, raptor perches may be desirable. If songbirds are more likely targets of the raptors, then residual trees may not be appropriate.

Dead standing trees may have little visual appeal, but there are many bird species that depend on them for nesting and feeding sites. You are encouraged to leave dead trees in harvest areas.

Before harvesting begins, carefully plan a system of skid trails, log landings, and haul roads. Locate landings out of public view behind vegetation screens or hills. Minimize the length of trails and roads and the number of stream crossings to minimize soil erosion and soil compaction. Soil disturbance is unattractive and a potential source of water pollution. Compacted areas also may not regenerate with trees or other vegetation as readily as less disturbed areas.

When clearing roadways, avoid bulldozing trees into large piles that will require many years to decay. Instead, use as much wood as possible for products. Push woody debris into depressions (not wetlands) or disperse it in the woodland. When bulldozing stumps from a roadway, push them off to the side so they remain upright. They appear more natural in this position. Construct and use roads so as to minimize erosion by following guidelines in Chapter 9.

Keep mud off public roads by harvesting when the ground is dry or frozen. As an alternative, maintain a hard surface or provide clean fill (gravel or wood chips) on the haul road for about 200 feet before the highway entrance.

During selection harvests, control the direction of fall so that damage to residual trees is minimized and the downed tree is positioned for skidding. This minimizes scars on the bark of residual trees. Leave bumper trees standing along roads and skid trails to protect nearby trees of better quality that will be left standing. Bumper trees can be cut last or left until the next harvest.

Woody debris left in a woodland can be unsightly. Remove as much wood as possible for products. Cut stumps low and lop slash so that it is no more than two to four feet above ground level. Woody debris decays more quickly when it is near the ground where humidity is relatively high. It also is concealed more quickly by new trees.

Unusable logs, limbs, and bark often accumulate at log landings. To reduce the amount of this debris, trim as much unusable wood from the trees as possible before logs are skidded to a landing. When logging has been completed, burn, bury, or disperse residual woody debris and reseed landings.

Be sure to clean up refuse and discarded equipment from the harvesting area.

REGENERATION

Plan timber harvests to encourage natural regeneration or to prepare sites for artificial regeneration. Choose a harvest system (see Chapter 4) that is compatible with the tree species you want to regenerate (see Chapter 6). Rapidly growing young trees quickly conceal logging debris and enhance aesthetics.

Plant trees in straight rows only when necessary for maintenance purposes; otherwise plant at random spacings for a more natural appearance. When practical, plant mixtures of tree species rather than single species to mimic nature.

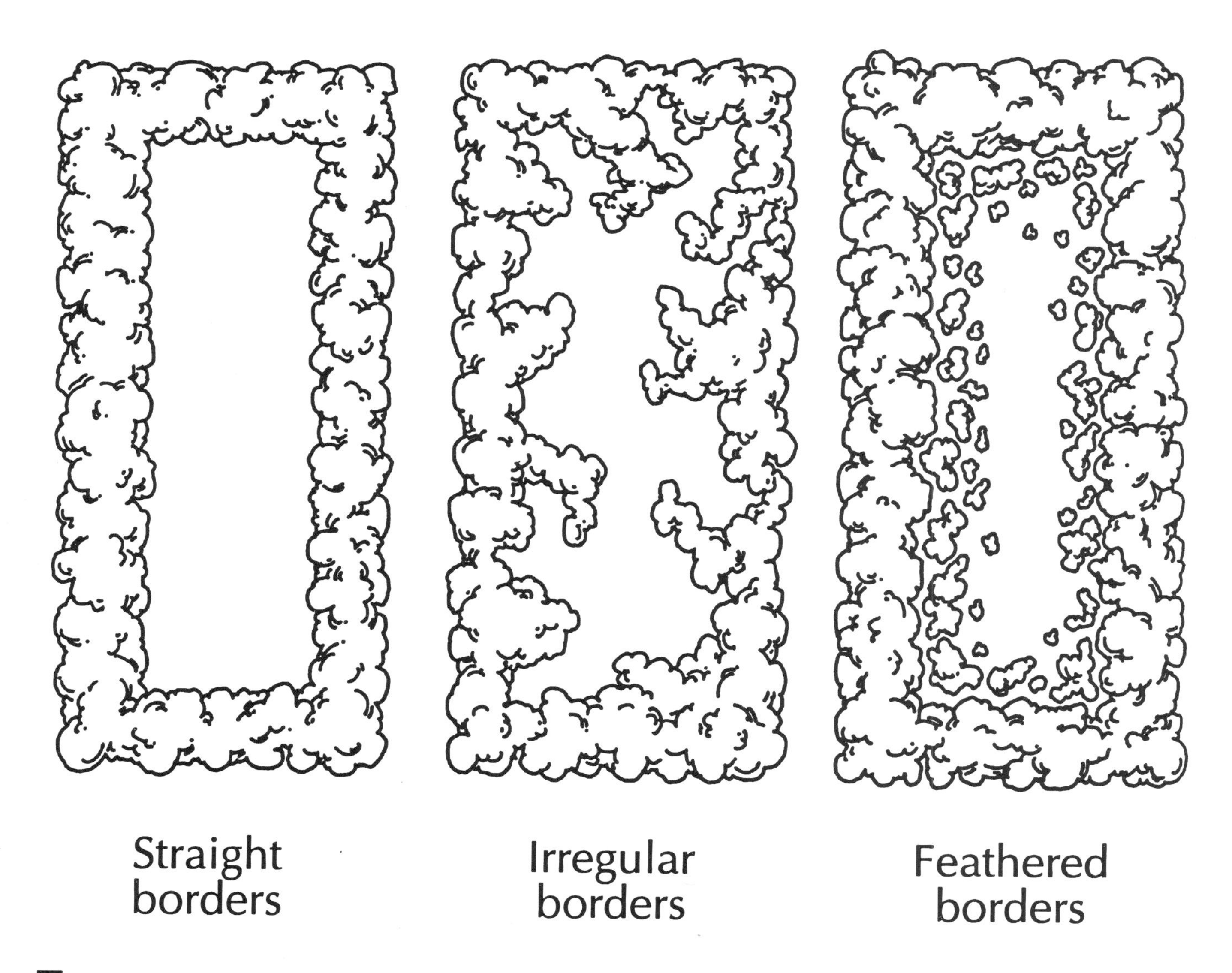

Fig. 67. Irregular or feathered borders are more appealing than straight borders around clearcuts.

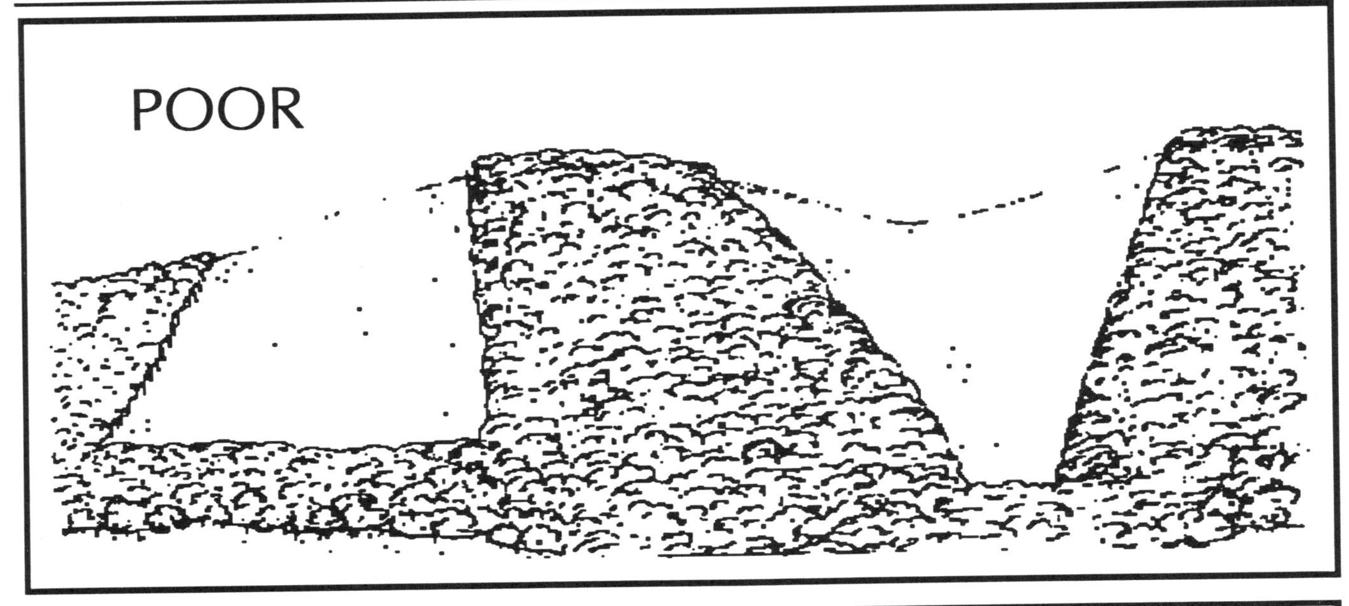

Fig. 68. In rolling topography, design clearcuts to blend into the landscape.

Site preparation may be necessary to control competing vegetation or prepare a seedbed. Controlled burning, mechanical scarification, or herbicides may be recommended. Practices that can be conducted in the spring just before greenup, in the fall just before leaves turn color, or during the dormant season are less visually obtrusive than summer treatments because they do not result in unsightly dead leaves on trees and shrubs.

WOODLAND IMPROVEMENT PRACTICES

Thinning, culling, weed-tree removal, and pruning are common management practices aimed at changing tree species composition, improving tree quality, and increasing tree growth rates (see Chapter 5). During these activities, harvest and use as much wood as economically feasible to minimize the amount of woody debris left on the forest floor. Cut up tree tops and pruned branches so they lie as close to the ground as possible.

If you need to kill unusable trees, minimize visual impact by killing them standing. Then they will fall down over a period of years. In contrast, if a large number of trees are felled at one time and left on the ground, they may form a temporarily impenetrable mass of debris. Deciduous trees that are killed standing will be less noticeable when killed during the dormant season. Deciduous trees killed during the growing season will retain dead, brown leaves for several months.

WOODLAND PROTECTION

As described in Chapter 7, woodlands need protection from fire, insects, and disease. Large numbers of trees damaged or killed by these agents are unsightly. Protect your woodland against these pests by following the recommendations in Chapters 6 and 7.

Suggested Reference

Jones, G. T. 1993. *A Guide to Logging Aesthetics (NRAES-60).* Northeast Regional Agricultural Engineering Service, Cooperative Extension Service, 152 Riley-Robb Hall, Ithaca, NY 14853-5701. 30 pp.

12

FEDERAL INCOME TAXES

This chapter describes how to handle many common woodland management transactions for federal income tax purposes. It includes tax information regarding purchases of land, timber, and equipment; tree planting expenses; management expenses; cost-share payments; timber sales; casualty losses; thefts; and condemnations.

Income and expenses associated with woodland management are subject to federal and state income tax regulations. Knowing how to report these transactions will enable you to pay only the tax that is due and maximize your financial returns from forestry investments.

Tax regulations change frequently. Consult the Internal Revenue Service (IRS) or your tax advisor for information on specific situations or to learn about recent changes in tax regulations.

DEFINING YOUR OPERATION

How you treat woodland management expenses and income for tax purposes depends on the purpose for which woodland is owned (personal use, investment, or business) and the type of taxpayer (individual or corporation). Before you consider specific tax situations, you must identify where your operation falls within these categories.

Purpose for Owning the Woodland

A woodland may be owned for three basic purposes: personal use, investment, or business.

Personal Use

Property not used to produce income (e.g., woodland acquired for a homesite or for recreational purposes) is held for personal use. You might expect to sell it later for more than you paid for it, but your primary purpose for acquiring it is to build a home or to use it for recreation.

Investment

If you own woodland primarily for the production of income, but have relatively infrequent business activities, it may be an investment. For example, if you own woodland to produce income from timber, but sell timber at intervals of five years or more, this most likely would be considered investment property.

Business

Property is held for a business if there are relatively frequent activities and production of income. If your woodland is a principal source of income and you have one or more timber sales annually, you may be operating a timber business. There are no clear guidelines to distinguish an investment from a business other than the relative frequency of business activities.

Your role in a business may be either **active** or **passive.** This distinction is important, because your classification affects how and when you can deduct expenses.

Active Participation in a Business

You are actively participating in a business if you "materially participate" on a year-round, regular, continuous, and substantial basis. For tax purposes this is the most advantageous category for woodland owners.

The amount of time spent on a business can serve as evidence of material participation. You are assumed to be materially participating if you meet any of the following criteria:

- You participate in the business more than five hundred hours during the tax year.
- Your participation in the business constitutes substantially all of the participation (including that of all other individuals) for the tax year.
- You participate in the business for more than one hundred hours during the tax year and no other individual participates more.
- Your aggregate participation in all of your "significant participation activities," including your timber business, exceeds five hundred hours during the tax year. (A business is a "significant participation activity" if it is a trade or business in which you participate for more than one hundred hours during the tax year. Thus, you could qualify under this test even if another individual who co-owns forest property with you participates in its operation more than you do during the tax year in question.)
- You materially participated in the business for any five of the preceding ten tax years. (Pre-1987 tax years count only if you meet the five hundred-hour test in those years.)
- All the facts and circumstances of the situation indicate that you and your spouse participated in the business on a regular, continuous, and substantial basis during the tax year.

Passive Participation in a Business

You are a passive participant in a business if you do not materially participate. Any rental activity is a passive activity whether or not you materially participate.

For purposes of determining whether the material participation requirement has been met, both you and your spouse are treated as one taxpayer. It does not matter whether your spouse owns an interest in the property, or whether you file joint or separate tax returns.

Type of Taxpayer

There are two basic types of taxpayers: individuals and corporations.

An individual engaged in a business as a sole proprietor reports most income and all expenses on either Schedule C or Schedule F of IRS Form 1040.

Corporations may be one of two types. Those electing to be taxed like a **partnership** are referred to as **S corporations.** Those not making the election are **C corporations.** Partnerships and S corporations do not pay income taxes, but must file information returns. Their costs and incomes are passed through to individual taxpayers who report their portion of partnership or S corporation income (loss) on IRS Form 1040, Schedule E.

EXPENSES

As a general rule, ordinary and necessary expenses incurred for the production of income are **deductible** (i.e., can be used to reduce taxable income). How and when deductions are claimed depends on the type of expense. Expenses can be classified into three categories: capital expenses, operating expenses, and sale-related expenses.

Capital Expenses

A **capital expense** is money spent to acquire a **capital asset** (e.g., real estate or equipment) or to make improvements that increase a capital asset's value. Examples include purchases of land, timber, buildings, machinery, and equipment having a useful life of more than 1 year; major repairs that prolong the life of machinery and equipment; acquisitions of rights-of-way extending more than one year; expenditures for tree planting; and construction of bridges, permanent roads, and firebreaks.

Usually you cannot deduct capital expenses in their entirety in the year they are incurred. Instead, record the **original basis** for the asset acquired in a special account called a **capital account** and deduct it from future earnings or reclaim it through depletion, depreciation, or amortization, depending on the type of asset. The original basis in your capital account may need adjustment from time to time to reflect additional expenses that you capitalize or expenses that you reclaim. Procedures for determining the original basis and setting up a capital account are discussed below.

Original Basis

Your original basis for a capital asset depends on how you acquire it.

If you **buy** an asset, your original basis is the acquisition cost. The acquisition cost includes the asset's **fair market value** plus any additional costs related to the purchase such as a land survey, forest inventory, and attorney fees. Fair market value is the price at which property would change hands between a buyer and a seller, neither being required to buy or sell, and both having reasonable knowledge of all of the necessary facts.

The original basis for an **inherited** asset is the asset's fair market value on the date the donor died or on the alternate estate valuation date provided by federal estate tax law. This fair market value is often higher than the donor's adjusted basis, resulting in a stepped-up basis for the heir. Upon sale of the asset, the heir may benefit from a substantial tax savings by deducting the stepped-up basis from the sale price to determine taxable income.

For assets received by **gift**, the original basis is the donor's adjusted basis at the time of the gift, plus that portion of the gift tax equal to the difference between the donor's adjusted basis and the gift's fair market value at the time of the gift.

Allocating Original Basis

When buying woodland, you may simultaneously acquire several assets (e.g., land, timber, buildings). Allocate the basis among the assets according to their respective fair market values so that when a separable asset such as timber is sold or destroyed, its basis can be deducted from its sale revenue or other compensation to reduce taxable income (Example 1).

♦ **Example 1.** You bought 100 acres of woodland in 1991. The fair market value was $30,000, but you also paid $1,000 for a boundary survey, $200 for a title search, and $800 for a timber inventory. Therefore, your total acquisition cost was $32,000. The timber inventory and subsequent valuation in 1991 showed that the tract contained 1,800 cords of merchantable wood with a fair market value of $9 per cord or $16,200 total, and 10 acres of premerchantable young growth valued at $30 per acre, or a total of $300. The fair market value of the bare land thus was $135 per acre for a total of $13,500.

Figure your original basis for each asset by determining the proportion of the total fair market value represented by each and multiplying this ratio by the total acquisition cost. Thus, the fair market value of the merchantable timber divided by the total fair market value of the property ($16,200/$30,000 = 0.54) multiplied by the total acquisition cost ($32,000) gives an original basis of $17,280 for the merchantable timber. The original basis for each asset is given below.

ASSET	FAIR MARKET VALUE (FMV, $)	PROPORTION OF TOTAL FMV	ORIGINAL BASIS ($)
Merchantable Timber	16,200	0.54	17,280
Young Growth	300	0.01	320
Land	13,500	0.45	14,400
TOTAL	30,000	1.00	32,000

If you acquired woodland several years ago, it may be possible to determine the fair market value of the timber when the property was acquired by having a forester inventory and appraise the timber, then deduct the volume and value that have grown since you acquired it.

Establishing Capital Accounts

Capital accounts are used to record the original basis of each of your separable capital assets. A good way to organize capital accounts is to set up separate accounts for land, depreciable land improvements, timber, and equipment, then divide each account into subaccounts for separable capital assets.

Land Account

Assets that can be placed in the land account include the land and permanent earthwork improvements (e.g., clearing, grading, and ditching of permanent roads; land leveling; impoundments). These earthworks are nondepreciable and their costs can be recovered only when you sell or otherwise dispose of the land.

Depreciable Land Improvements Account

Depreciable assets are those held for use in a business or for the production of income that have a determinable useful life and wear out, decay, become used up, become obsolete, or lose value from natural causes. Depreciable land improvements include bridges, culverts, temporary roads, gravel for roads, fences, and nonpermanent structures and improvements. By establishing separate accounts for depreciable land improvements, you can recover their costs over a period of years according to depreciation schedules established by the IRS rather than waiting to recover their costs when the property is sold or otherwise disposed of.

Timber Account

A timber account commonly has three subaccounts: merchantable timber, young growth (naturally regenerated trees of premerchantable size), and plantations (planted trees of premerchantable size). Values in the young growth and plantation accounts are transferred to the merchantable timber account when the trees reach merchantable size.

There should be a different set of timber accounts for each **block** of timber. Block boundaries may be based on geographic boundaries (e.g., stands or tracts), tree species, timber products, accessibility, or other timber characteristics. If you own a small woodland, treat the whole woodland as one block. If you own a large woodland (e.g., more than 1,000 acres) it may be advantageous to establish block boundaries that allow you to assign your highest costs to assets that will be sold soon, thus enabling you to recover costs quickly.

Each timber subaccount should include two entries, one showing the quantity of timber and the other showing its original basis. Merchantable timber usually is measured in cords or thousand board feet. Premerchantable timber usually is measured in acres.

If you do not keep a separate merchantable timber account, you will not be able to recover the timber's basis when it is sold. In effect, you have lumped the timber with the land and will be able to recover the basis in your timber only when you sell the land.

Adjust your timber account whenever you add to the basis or claim deductions for depletion, amortization, and sale or other disposal of the asset (Example 2).

♦ **Example 2.** In 1992 you hired a forester to remeasure the timber you bought in 1991 (see Example 1). You found that your 90 acres of merchantable timber had grown by 45 cords. You therefore added 45 cords to the quantity of timber in your merchantable timber account. The young growth on 10 acres had reached merchantable size and was estimated to contain 50 cords. You transferred the original basis and current volume in the young growth account to the merchantable timber account. In 1992 you purchased 50 acres of woodland next to your original tract. It contained 1,000 cords of timber with an original basis of $9,000, which you also added to your account. The adjusted merchantable timber account in 1992 is as follows:

TRANSACTIONS	QUANTITY (CORDS)	ADJUSTED BASIS ($)
Merchantable timber in 1991	1,800	17,280
Addition for growth (1991-1992)	45	0
Young growth that became merchantable in 1992	50	320
Timber acquired in 1992	1,000	9,000
Net quantity and basis in 1992	2,895	26,600

Report your original basis for timber on IRS Form T (Timber) Forest Industries Schedules when you first acquire timber and report adjustments in the timber account whenever you claim a depletion allowance or dispose of timber.

Equipment Account

This account should have subaccounts showing costs for each item or class of items of equipment, including power saws, tractors, trucks, and tree-planting machines. Increase the original basis of equipment by any amount spent for major repair or reconstruction that significantly increases the value or prolongs the life of the equipment. You recover these costs through depreciation.

Reclaiming Capital Expenses

Capital expenses are reclaimed through depreciation, amortization, depletion, or deduction at the time the capital asset is disposed of. **Depreciation**, the process of deducting the original basis of an asset over a period of years according to IRS schedules, applies mainly to equipment and buildings. **Amortization**, which applies to tree planting expenses, is much like depreciation in that expenses are deducted over a period of years according to an IRS schedule. **Depletion** is the process of deducting the basis for timber from timber sale or other income. The costs for some capital assets, notably land, are recovered only when the asset is sold or otherwise disposed of.

Depreciation and the Section 179 Deduction

You may recover the cost of machinery, equipment, and certain land improvements through annual depreciation deductions.

If your timber operation is a business in which you materially participate, you may be able to deduct up to $10,000 of the property's cost, in addition to the normal depreciation deduction, in the year it was acquired. This is an Internal Revenue Code **Section 179** deduction. You must elect to claim the deduction in the tax year during which the property is first placed in service or on an amended return filed within the time prescribed by law for that year.

The $10,000 deduction is reduced $1 for each $1 over $200,000 you invest during the tax year in Section 179 property. For example, if you purchase $206,000 worth of Section 179 property, only $4,000 qualifies for this deduction; the remainder must be depreciated.

To depreciate property not recovered by a Section 179 deduction, you must use the Modified Accelerated Cost Recovery System (MACRS) for most tangible property (property that can be seen or touched, such as buildings and equipment) placed in service after 1986. You cannot change to MACRS for property placed in service prior to 1987 that is being depreciated under another method such as the Accelerated Cost Recovery System.

Under MACRS, each property class is assigned a class life that determines the depreciation recovery period. Recovery periods under this system are 3, 5, 7, 10, 15, and 20 years for most tangible property, but they are 25.5 years for residential rental property and 31.5 years for nonresidential real property (Table 12).

Intangible property that has a determinable useful life (e.g., patents, copyrights, and franchises) can be depreciated using the straight line method.

Calculating depreciation is a complex process. For further information, please refer to IRS Publication 534 (*Depreciation*). Depreciation is reported on IRS Form 4562.

Reforestation Amortization and Investment Tax Credit

Expenses incurred for establishing timber stands by planting or seeding should be **capitalized** (recorded in a capital account). Up to $10,000 ($5,000 for married persons filing separately) of these expenses annually are eligible for **amortization** and a 10 percent investment tax credit. Record expenses that exceed $10,000 in your plantation subaccount. When the trees reach merchantable size, transfer those expenses to the merchantable timber account.

For an expense to be eligible for amortization and the tax credit, it must be incurred for forested property in the United States that is used in the commercial production of timber products and is greater than one acre. You may amortize the direct costs of stand establishment, such as site preparation, seed or seedlings, labor, tools, and depreciation of equipment used in planting or seeding.

Expenses for improvement of established timber stands (e.g., thinning, pruning, or culling) do not qualify for amortization or the tax credit, nor do expenses for planting Christmas trees, shelterbelts, windbreaks, or nut trees. Reforestation expenses reimbursed under a public cost-share program are not eligible for amortization or the tax credit, unless those reimbursed expenses are included in taxable income. In the latter case, the total reforestation expense is eligible for amortization and the tax credit, subject to the $10,000 limitation.

Individuals, estates, partnerships, and corporations are eligible for the amortization and tax credit. Trusts are not eligible for either. The $10,000 annual limit applies to both the partnership and each partner, and to both the S corporation and each shareholder. A partner's or shareholder's total annual amortizable reforestation expense from all sources cannot exceed $10,000.

Table 12. Recovery periods for forestry-related property under the modified accelerated cost recovery system (MACRS).

MACRS Recovery Period (YEARS)	TYPE OF PROPERTY
3	Over-the-road (semi) tractors.
5	Automobiles, light general-purpose trucks, heavy general-purpose trucks, logging machinery and equipment and road-building equipment used by logging and sawmill operators and pulp manufacturers for their own account, portable sawmill equipment, computers, typewriters, calculators, and assets used in construction.
7	Machinery and equipment used in agriculture or in the operation of nurseries, greenhouses, and fur farms; agricultural fences; machinery and equipment used in permanent sawmills; assets used in production of plywood, hardboard, flooring, veneers, furniture, treated poles, timber, pulp, and paper; assets used for modification or remanufacture of paper, paperboard, and pulp products; office furniture and fixtures not a structural component of a building; any property that does not have a class life and is not otherwise classified under Internal Revenue Code Sections 168(e) (2) or (3).
10	Single purpose agricultural or horticultural structures; vessels, barges, and similar water transportation equipment, except those used in marine construction; assets used in provision of electric power by generation and distribution.
15	Land improvements for Internal Revenue Code Sections 1245 and 1250 property, including sidewalks, roads, canals, waterways, drainage facilities, sewers, wharves, docks, bridges, fences, and landscaping shrubbery; water transportation assets used in commercial and contract carrying of freight and passengers.
20	Farm buildings (except single-purpose agricultural or horticultural structures).
27.5	Residential rental property.
31.5	Nonresidential real property.

To amortize reforestation expenses, deduct 1/14 of the expense in the year incurred, 1/7 in each of the next six tax years, and the remaining 1/14 in the eighth tax year.

The 10 percent tax credit applies if the trees live more than 7 years. If you claim the full credit, you must reduce the amortizable basis of the property by half the amount of the credit claimed. For example, when you take a 10 percent tax credit, you must reduce the amortizable basis by 5 percent, leaving 95 percent of the reforestation expense to be amortized (Example 3). Expenses not amortized can be capitalized and recovered when the timber is sold. If you have a passive role in a timber business, the investment tax credit can offset taxes payable only on passive income.

♦ **Example 3.** It cost you $7,500 to plant 50 acres of trees. You received 50 percent cost-sharing and decided to include cost-share payments in your income. You chose to claim the full 10 percent tax credit; therefore, you may amortize 95 percent of $7,500, or $7,125. The remaining 5 percent ($375) must be capitalized and recovered when the trees are sold. For the tax year in which you incur planting expenses, you deduct 1/14 of $7,125, or $508.93. In each of the next six tax years, you deduct 1/7 of $7,125, or $1,017.86. The remaining 1/14 of $7,125, or $508.93, is deducted in the eighth tax year. In addition, you claim a $750 tax credit (10 percent of $7,500) in the year reforestation expenses were incurred.

If you manage your woodland as an investment, show the amortization deduction on the line for "adjustments to income" on the front page of IRS Form 1040 by writing "reforestation" and the amount on that line. You need not report amortization as an itemized deduction. If you manage your woodland as a business, report the amortization deduction on the "other expense" line of Schedule C of IRS Form 1040. Farmers can report amortization deductions on the "other" expense line of Schedule F of IRS Form 1040.

If you plant trees in more than one year, you must maintain separate amortization schedules for each plantation. Amortization must be elected on a tax return filed on time, including extensions. The election cannot be made on an amended tax return. Once the election is made, however, missed amortization deductions may be reported on amended returns.

The investment tax credit is claimed on IRS Form 3468 in the section for nonrecovery property. If you have an active role in your timber business, the tax credit can be applied to taxes associated with income from any source. If you have a passive role in your timber business, the tax credit can offset tax payable only on passive income. Tax credits you cannot use during a tax year may be carried forward for use in future years, but may not be taken in the year that you dispose of your entire ownership interest.

If the trees are disposed of within 7 years, part or all of the tax credit is subject to recapture (i.e., you will have to pay back to the IRS part or all of the tax credit). If the trees are disposed of within 10 years, part or all of the amortization is subject to recapture.

Operating Expenses and Carrying Charges

Operating expenses include what you spend for tools of short life (usually 1 year or less) or low cost (e.g., axe, hand saw); equipment operation and maintenance; salaries or other compensation (e.g., hired labor, consulting forester, lawyer, accountant); travel expenses directly related to the income potential of the property; rental of land, equipment, or other property; expenses for fire, insect, and disease protection; and expenses for precommercial thinning and timber stand improvement (e.g., labor, equipment, materials) after the stand is established. **Carrying charges** include property taxes, interest payments on loans, insurance premiums (e.g., fire, windstorm, theft, general liability, workers' compensation), and certain other expenses related to property development and operation.

As a general rule, ordinary and necessary expenses for managing, maintaining, and conserving woodland may be deducted if you are growing timber for profit. It is presumed that an activity was engaged in for profit if there was net income in 3 or more years of a consecutive 5-year period. If there was no profit in 3 of 5 consecutive years, then other facts and circumstances are considered in determining whether the activity was engaged in for profit. Fortunately for timber growers, profit includes appreciation in the value of assets, which occurs as timber grows over time. If audited, however, you may need to prove by a financial analysis or other means that you expect to make a profit as a result of your timber-related expenses.

It is generally to your advantage to currently deduct operating expenses and carrying charges, but you may need to capitalize them if you do not have sufficient income to offset the expense deduction. As a general rule only carrying charges may be capitalized. However, there are exceptions that permit capitalization of some operating expenses, and many timber-growing expenses fit within the exceptions.

You elect to capitalize expenses by including a statement with your tax return for the year the election takes effect describing the expenses you are capitalizing.

The extent to which you are permitted to currently deduct operating expenses and carrying charges depends on whether the expenses are for a business in which you are an active or passive participant or for an investment.

Active Participant in a Business

If you have an active role in a business, you may fully deduct all management expenses, taxes, and interest each year from income from any source. Management expenses include all operating expenses and carrying charges except taxes and interest. If deductions from the business exceed gross income from all sources for the taxable year, the excess net operating loss generally may be carried back to the three preceding tax years and, if necessary, carried forward to the next succeeding fifteen tax years. You may instead elect to capitalize these timber-related expenses and recover them when the timber is sold.

If your timber operation is incidental to farming, list your deductible timber expenses on Schedule F of IRS Form 1040 on the line for "other" expenses. If your timber operation is a separate business, or is incidental to a nonfarm business, report your timber deductions on Schedule C of IRS Form 1040 on the line for "other" expenses. On either schedule list your deductions individually. There are separate lines for tax and interest deductions.

Passive Participant in a Business

If you have a passive role in a business, you may currently deduct management expenses, taxes, and interest only to the extent that, when aggregated with all other expenses of your passive business or investment activities, they do not exceed the income from such activities. Passive losses cannot be deducted from income arising from any trade or business in which you materially participate, from wage or salary income, or from investment income. An exception is that C corporations that are not classified as closely held or as personal service corporations may deduct in the current tax year management expenses, taxes, and interest expenses associated with passive timber ownership from income from any source without limit.

Expenses that cannot be deducted during the year incurred may be carried forward to years in which you either realize passive income or else dispose of the entire property that gave rise to the passive loss. You may instead elect to capitalize expenses and deduct them from income realized when the property is sold.

Passive loss rules are fully effective with respect to losses arising from passive activities in which you acquired an ownership share after October 22, 1986. Losses arising from passive activities in which an ownership share was held on or before that date are subject to a five-year phase-in of the rules. Passive loss rules became fully effective in 1991 (see IRS Publication 925, *Passive Activity and At-Risk Rules*).

Investment

Both corporate and noncorporate woodland owners generally may deduct management expenses relating to timber held as an investment against income from any source. However, for noncorporate taxpayers, management expenses are **miscellaneous itemized deductions** and may be deducted only to the extent that, when aggregated with all other miscellaneous itemized deductions, the total exceeds 2 percent of adjusted gross income. The portion of miscellaneous itemized deductions that falls below the 2 percent floor is permanently lost. If you capitalize rather than deduct management expenses, they may not also be counted toward the 2 percent floor on miscellaneous itemized deductions.

Both corporate and noncorporate taxpayers may fully deduct property and other deductible taxes (e.g., severance and yield taxes) each year against income from any source. Deductions for taxes are not miscellaneous itemized deductions and so are not subject to the 2 percent floor. You may elect to capitalize taxes (except severance and yield taxes) and recover them upon sale of the timber, rather than deduct them in the year paid.

Corporate taxpayers may deduct an unlimited amount of investment interest expense against income from any source. Noncorporate taxpayers may deduct investment interest expense from all investments up to the total net investment income from all investments. You may elect to capitalize all or part of the interest paid instead of currently deducting it.

List investment expenses on Schedule A of IRS Form 1040 on the appropriate line for each type of deduction. If you do not itemize deductions for the year or elect to capitalize these expenses, the expenses are lost for tax purposes.

Sale-Related Expenses

When selling timber, you may deduct sale-related expenses from timber sale income to determine net taxable income. Deductible expenses include advertising, timber cruising, marking, scaling, and hiring a consulting forester or lawyer. These expenses are deductible in the year of sale regardless of your purpose for holding timber or the type of taxpayer. The forms on which you report these expenses, however, vary by the type of taxpayer and method of sale as described below under "Timber Sale Income."

COST-SHARE PAYMENTS

How you treat cost-share payments depends on whether they were received for timber stand improvement or reforestation.

Timber Stand Improvement

If you receive a cost-share payment for **timber stand improvement** (e.g., weed or brush control after stand establishment, pruning, culling, precommercial thinning), you must report it as income. You then may deduct or capitalize the full cost of the practice as you would other management expenses.

Reforestation

Some or all of a cost-share payment for **reforestation** may be excluded from taxable income if two provisions are met: (1) the Secretary of Agriculture determines that payment is primarily for the purpose of conserving soil and water, protecting or restoring the environment, improving forests, or providing wildlife habitat; and (2) the Secretary of the Treasury or the Secretary's delegate determines that payment does not substantially increase your annual income from the property.

Cost-share payments for tree planting and recurring annual payments under the Conservation Reserve Program have not been approved by the Secretary of Agriculture for the exclusion. You must include these payments in income. However, you can amortize tree planting costs (including the portion reimbursed by a cost-share payment), or you may be able to currently deduct them if you are in the business of farming.

Cost-share payments for tree planting under the Forestry Incentives Program and the Agricultural Conservation Program have been approved by the Secretary of Agriculture for the exclusion. However, the total amount you may exclude is subject to limits defined by the Secretary of the Treasury. You also have the option to include cost-share payments in income if you choose.

Include Cost-Share Payment

You may include the full cost-share payment for tree planting in your taxable income. Then you may amortize and claim an investment tax credit on the total cost (up to $10,000), including the share you pay and the cost-share payment.

Exclude Cost-Share Payment

You may exclude from income your cost-share payment for tree planting, subject to certain limits. You may not exclude cost-share payments from income when those payments substantially increase your income. The maximum cost-share payment that may be excluded is the present fair market value of the right to receive from the affected acreage annual income equal to the greater of these two amounts: (1) 10 percent of the average annual income from the affected acreage prior to reforestation or other qualifying activity (Example 4) or (2) $2.50 times the number of affected acres (Example 5). Amounts included in income are eligible for a tax credit and amortization, but amounts excluded are not.

IRS regulations do not describe how to determine the fair market value of the right to receive annual income. A common method for determining the value of annual income over an indefinitely long period is to divide the annual income by the annual interest rate that you could expect to earn in your next best investment alternative for the money you invested in the woodland over a similar time period.

Prior average annual income is defined as the average of gross receipts from the affected acreage for the 3 tax years immediately preceding reforestation.

- **Example 4.** Assume that you harvested $12,000 worth of timber from 20 acres and then within 3 years you reforested the tract, receiving $1,500 in cost-share payments for part of the reforestation expense. Assume the interest rate you could earn in your next best investment alternative over a time period similar to your timber investment is 8 percent. Your prior average annual income is your gross income ($12,000) divided by 3 years, or $4,000 per year. The fair market value of the right to receive this prior average annual income is your annual income ($4,000) divided by your interest rate (0.08), or $50,000. You may exclude 10 percent ($5,000) of this amount from taxable income. Since your cost-share payment was $1,500, far less than the $5,000 maximum exclusion, you may elect to exclude the entire cost-share payment.

- **Example 5.** If you had no income in the previous 3 years from the land where you planted trees, then the maximum amount that you may exclude from gross income is the fair market value of the right to receive $2.50 times the number of affected acres. Assume the interest you could earn on your next best investment alternative is 8 percent and you received $1,500 in cost-share payments for planting 20 acres of trees. The fair market value of the right to receive $2.50 is $2.50 divided by your interest rate (0.08). This value is $31.25 per acre, or $625 for 20 acres. You may exclude from taxable income $625 in cost-share payments. The remaining $875 in cost-share payments must be included in income, but you may claim a tax credit and amortization on this amount.

Since no amortization, investment tax credit, or addition to basis may be made with respect to cost-share payments excluded from income, some taxpayers are better off financially if they include cost-share payments in income. The decision to include or exclude, when that option is available, should be based on which alternative maximizes tax savings from amortization and the tax credit, discounted to the present time using an interest rate that is the highest you could earn in your next best investment alternative over the 8-year amortization period.

If you exclude any portion of your cost-share payment, include a statement with your tax return indicating the total reforestation cost, the amount of the cost-share payment, date received, purpose of the payment, and the amount excluded.

Payments excluded from taxable income may be subject to recapture as ordinary income if trees are disposed of within 10 years of establishment. The amount recaptured will be the lesser of (1) the amount of gain from the disposal, or (2) the amount of cost-share payment excluded. The proportion subject to recapture declines 10 percent per year for each year the property is held beyond 10 years.

TIMBER SALE INCOME

Net income from timber sales is taxable. When you sell standing timber, you must determine the amount as well as the type of income for federal income tax purposes. This section describes how to handle income. However, if your timber sale results in a net loss, you may apply the same principles in figuring your taxes.

Determining Amount of Income

Net income from a timber sale is the gross income minus the sale-related expenses (e.g., advertising, timber cruising, marking, scaling, hiring a consulting forester or lawyer) and the depletion allowance for the harvested timber.

The **depletion allowance** is a portion of the adjusted basis. It is the cost you incurred to produce the timber that is sold. If you acquire standing timber and later sell the entire stand in a single transaction, your depletion allowance would be your entire adjusted basis in the timber. If you acquire standing timber, then sell portions of it in several transactions over a period of years, you need to calculate the depletion allowance for each portion of the timber sold. To do so, first divide the total presale quantity of timber in the block by the adjusted basis of the timber to get the **depletion unit**, or cost per unit of timber. Next multiply the depletion unit by the quantity of timber sold to get the depletion allowance for that timber (Example 6).

♦ **Example 6.** In 1992 you sold 1,000 cords of timber from your 150-acre tract (see Example 2). First divide the adjusted cost basis ($26,600) by the total quantity of timber on the property (2,895 cords) to get the depletion unit ($9.19 per cord). Multiply the depletion unit ($9.19) by the number of units sold (1,000 cords) to get the depletion allowance that can be deducted ($9,190).

TRANSACTIONS	QUANTITY (CORDS)	ADJUSTED BASIS ($)
Merchantable timber in 1992 before harvest	2,895	26,600
Timber sold in 1992	(1,000)	- 9,190
Net quantity and adjusted basis at the end of 1992	1,895	17,410

Following this sale you still have 1,895 cords of timber remaining on the property with an adjusted cost basis of $17,410. In the future when you sell more timber, you will once again need to calculate the depletion allowance for the portion sold. You may need to remeasure your timber to estimate the total quantity of timber on the property at the time of that next sale.

Report your depletion allowance on Schedule F of IRS Form T. Profit and sale-related expenses are reported on Schedule C of IRS Form T.

You may not claim a depletion allowance for timber harvested for personal use, such as firewood for your home. Do not adjust the dollar amount in your merchantable timber account when you cut timber for personal use. However, if you cut very much timber, you may need to adjust the quantity of timber shown in the account to reflect the lower quantity available for commercial sale.

Determining Type of Income

For income tax purposes you must determine whether your standing timber is a **capital asset** or a **noncapital** (ordinary) **asset**. This distinction determines whether your timber sale income is ordinary or capital income and subsequently how you report the transaction.

Even though the tax rate differential between ordinary income and net capital gains was eliminated by the 1986 Tax Reform Act, it still may be important to report timber sale income as capital gain. For example, only $3,000 of capital losses may be offset against ordinary income, but there is no offset limitation against capital gains. In addition, if you are a sole proprietor or partner whose timber holdings are considered a business, you are subject to self-employment tax on ordinary income but not on capital gains income from the business. Also, Congress is considering reducing the tax rate on capital gains below the rate for ordinary income. If this occurs, it will be very advantageous to treat timber sale income as a capital gain.

Capital gains are categorized as long-term or short-term. To qualify for long-term capital gains, you must have owned timber (or the contract right to harvest timber) for more than one year prior to harvest. If you sell timber acquired by gift, both the donor's and your time of ownership may be counted toward the holding period. For inherited timber, no holding period is required in order to qualify for long-term capital gains status.

Whether your timber sale income is a capital gain depends on your method of timber sale and primary purpose for holding timber. You may sell timber in three ways: (1) on a lump sum basis, (2) on a sale-by-unit basis, or (3) by harvesting the timber yourself and selling the products.

Selling Timber on a Lump Sum Basis

When you sell timber for a fixed amount agreed upon in advance of the harvest (a lump sum sale), you may treat the income as a capital gain if you held the timber as an investment or for personal use and not for use in your business or for sale to customers (Example 7). The income is a long-term capital gain if the minimum holding period described above has been met.

- **Example 7.** In 1992 you sold 1,000 cords of timber from your 150-acre tract (see Example 6) for $11,000 payable before harvesting. A consulting forester charged $800 to cruise, mark, and sell the timber. Your depletion allowance for the sale was $9,190. You deduct your depletion allowance and sale expenses from the timber sale revenue, leaving a taxable long-term capital gain of $1,010.

Report lump sum timber sales for which proceeds qualify as capital gains on Schedule D of IRS Form 1040. If the long-term holding period has been met, enter the transaction in Part II. If the holding period has not been met, enter the information in Part I. If your income does not qualify for capital gains treatment because you held the timber for use in your business or for sale to customers, report the income on Schedule C or Schedule F of IRS Form 1040.

Selling Timber on a Sale-by-Unit Basis

Timber harvested under a contract that requires payment at a specified amount for each unit of timber harvested (e.g., $10 per cord), is a disposal with an economic interest retained. The income is considered a capital gain under Section 631(b) of the Internal Revenue Code whether the timber was held primarily for personal use, investment, or sale as part of a business. If the minimum holding period requirement has been met, it is a long-term capital gain.

Your income is calculated in the same manner as for a lump sum sale. Timber qualifying for capital gains status under Section 631(b) is Section 1231 property. Report gains and losses on IRS Form 4797. If aggregate Section 1231 gains exceed aggregate losses, the net gain is treated as a long-term capital gain and is transferred to Part II of Schedule D, IRS Form 1040. If the Section 1231 losses exceed gains, the net loss is treated as an ordinary loss and is transferred to Part II of IRS Form 4797.

Harvesting Timber, Then Selling Products

If you harvest standing timber and sell the logs or other products, report the income as ordinary income, unless you make a Section 631(a) election. You may make a Section 631(a) election only if you meet the minimum holding period requirement and you owned the timber (or the contract right to harvest the timber) on the first day of the tax year. If you make a Section 631(a) election, report the transaction in two parts:

1. Report as Section 631(a) income the difference between the depletion allowance for the timber that was harvested and its fair market value as standing timber on the first day of the tax year in which it was harvested. The timber must be valued as it existed on the first day of

the tax year regardless of any changes to it between that date and the harvest date.

2. Report as ordinary income the profit from harvesting the standing timber and converting it into sellable products. Profit equals gross income from sale of the products minus logging expenses and the fair market value of the standing timber on the first day of the tax year in which it was harvested (Example 8).

♦ **Example 8.** Assume you hired a logger to harvest 60,000 board feet of timber from a tract you purchased several years ago. Then you sold the harvested logs to a sawmill. Assume your depletion allowance for the standing timber was $1,200. Based on sales of comparable timber in the area, a forester estimated the fair market value of the standing timber on the first day of the year in which it was harvested as $4,800. Your logging expenses were $1,100. The sawmill paid you $6,000 for the logs. Since you had owned the timber longer than the 1-year minimum holding period, you may report part of your earnings as a long-term capital gain. You determine the income from harvesting separately from the income from selling the sawlogs as indicated on the chart (below).

TRANSACTION	AMOUNT ($)
1. Net income from harvesting timber:	
Fair market value of standing timber on first day of tax year	4,800
Depletion allowance	- 1,200
Long-term capital gain	3,600
2. Net income from selling sawlogs:	
Gross income from selling sawlogs	6,000
Fair market value of standing timber on first day of tax year	- 4,800
Logging expenses	- 1,100
Net ordinary income	100

You report a $3,600 long-term capital gain from selling standing timber. You also report ordinary income of $6,000 and ordinary expenses of $5,900 for an ordinary income of $100 from selling logs.

Elect to use Section 631(a) by providing information requested in items 44 through 51 on Schedule F of IRS Form T and by computing your taxes according to provisions of this section. If you elect to use Section 631(a) in a certain year, you must use it for all eligible timber you harvest that year and in all subsequent years. If you harvested timber under a Section 631(a) election before January 1, 1987, you may revoke this election once and reelect it once without permission from the IRS. The revocation may be advantageous if harvested timber is not sold in the same tax year in which it is harvested. Without the revocation, you will be taxed in the year of harvest on the standing timber's gain in value even though you earned nothing from sale of the products.

Report a gain or loss on the standing timber on IRS Form 4797 with other Section 1231 transactions. Report income from the sale of harvested products on Schedule C or Schedule F of IRS Form 1040. List the expense of timber harvested (fair market value used for computing gain) and the expenses of cutting and sale as "other" expenses on Schedule F or Schedule C of IRS Form 1040. Also attach Schedule F of IRS Form T and a description of how you estimated the fair market value.

CASUALTIES, THEFTS, AND CONDEMNATIONS

If part or all of your timber is destroyed, stolen, or condemned for public use, you may be entitled to deduct the loss on your income tax return. This section discusses timber losses for property held for investment or business use.

For information concerning losses to property held for personal use, including shade tree damage, refer to IRS Publication 547, *Nonbusiness Disasters, Casualties, and Thefts.*

Normal Loss

In all woodlands there is a "normal" level of tree mortality and damage resulting from natural causes such as insects, disease, drought, and competition among trees. For tax purposes, this normal loss is reflected in the quantity of timber shown in your timber accounts. Since the depletion unit is the quantity of timber divided by the adjusted basis, it reflects additions for growth and reductions for mortality. Normal losses increase the depletion unit, thereby reducing the taxable income when timber is sold. This process does not result in a separate deduction. To be deductible, losses must exceed normal levels.

Noncasualty Loss

Damage to or destruction of timber held for investment or business use is a deductible noncasualty loss if the loss exceeds a normal level, the cause is unusual and unexpected, and there is a closed and completed transaction associated with the loss. Such circumstances do not often apply to timber, but there is a case on record in which the IRS allowed a business to deduct losses caused by an insect attack that killed trees over a nine-month period.

Casualty Loss

Deductible timber losses generally must be caused by a **casualty**—an event that is sudden and unexpected or unusual rather than gradual deterioration through a steady force. The IRS interpretation of a casualty has emphasized suddenness. Casualties can result from fire, windstorm, ice storm, or hail. Timber losses from disease are not a casualty. Losses from insect attacks are seldom casualties, although a casualty loss was allowed for ornamental pine trees killed in a few days by southern pine beetles, which had not been active in the area for many years.

To qualify as a casualty, the damage must make timber unfit for use. An ice storm that breaks limbs and subsequently slows tree growth or leads to decay does not cause a casualty because the timber is still useful, albeit less valuable. Thus, you can not claim a casualty loss for the reduction in future profits resulting from damage that slows tree growth or reduces tree quality.

The IRS requires that you make every attempt to salvage timber that has been damaged but not made unfit for use. You cannot claim a casualty loss if the timber is salvageable. Keep records that show your attempts to salvage timber.

In the event of a casualty, the deductible loss is limited to your depletion allowance (adjusted basis) for that timber destroyed, less any insurance, salvage, or other compensation received. If your timber has no basis, you cannot claim a casualty loss. The depletion unit is based on the same units of measure that you use in your timber accounts (e.g., cords or thousand board feet). Figure the depletion allowance as you would for a sale (see Example 6). You may need to employ a consulting forester to determine the volume and value of timber destroyed.

Deduct a casualty loss in the year the casualty occurs. If you expect reimbursement from salvage, insurance, or other source but have not received payment, report as a casualty loss the amount of your depletion allowance that exceeds your expected reimbursement. If you later recover less than the amount you estimated, deduct the difference in the year in which you become certain that no more reimbursement or recovery can be expected. If you later recover more than the amount you estimated, you must include such excess as income on your return for the year received. Do not file an amended return for the year in which you claimed the original deduction.

Income from a timber salvage sale following a casualty is taxable if it exceeds the depletion allowance. Gain that exceeds the depletion allowance is tax deferred if it is reinvested in qualified replacement property as described below under "Postponing Gains from Involuntary Conversions."

You may claim a casualty loss for destruction of premerchantable plantation or young growth if you maintained separate plantation or young growth accounts and have allocated costs to such accounts. Figure the depletion allowance by dividing the adjusted basis in the account by the total acreage of plantation or young growth and then multiplying that depletion unit by the number of acres destroyed.

The ordinary and necessary expenses you incur to measure your loss (e.g., appraisal and timber cruising) are considered expenses in determining your tax liability. They are not part of the casualty loss. Corporations may fully deduct these expenses, but for individuals, they are **miscellaneous itemized deductions** and may be deducted only to the extent they exceed 2 percent of adjusted gross income.

Report casualty losses on IRS Form 4684, attaching a description of the entries. File a separate form for each event. Transfer losses from business or rental property to IRS Form 4797. Report losses from investments on Schedule A of IRS Form 1040. You also must file Schedules C, D, and F of IRS Form T if you are claiming a depletion allowance.

Theft

A loss from timber theft is limited to your depletion allowance for the timber stolen, less insurance or other compensation received. The quantity of timber used in determining the depletion unit is the quantity at the time the theft is discovered. If you later receive more compensation than you estimated as your deductible loss, report it in the year received as ordinary income.

If the timber thief is found and convicted, you may be awarded double or triple damages. Report the portion of the award that represents reimbursement for timber stolen as proceeds from an involuntary conversion. Report any additional damage award as "other income."

Deduct a theft loss in the year it is discovered. The ordinary and necessary expenses incurred in determining your theft loss are expenses in determining your tax liability. Report theft losses on the same forms used for casualty losses.

Condemnation

A **condemnation** is the lawful taking of private property by a government unit for public use without the owner's consent but with compensation. The tax consequences are the same if you sell property under threat of condemnation. First determine your basis as you would for a sale. Deduct expenses incurred to receive a condemnation award (e.g., attorney, appraisal, timber cruising) from the award to determine the net award to report on your tax return.

Condemnations may have complex tax consequences. Refer to IRS Publication 549 (*Condemnations and Business Casualties and Thefts*) for further information.

Postponing Gains from Involuntary Conversions

An **involuntary conversion** (exchange) occurs when your property is destroyed, stolen, condemned, or disposed of under the threat of condemnation and you receive compensation from salvage, insurance, or a condemnation award. You realize a gain on an involuntary conversion when the compensation is greater than your basis in the property. Your gain is not taxed in the year realized if, within the allowable replacement period, you purchase other property that is similar or related in service or use to the property converted (or you purchase the controlling interest in a corporation owning such property) at a cost that equals or exceeds the amount you received as compensation. For the loss of real property, including timber, the replacement period ends 3 years after the close of the first tax year in which any portion of the gain from the conversion is realized. The replacement period is 2 years for other property.

If you elect to defer reporting the gain, you must file a statement with your tax return stating that you are making the election and including all pertinent information concerning the conversion and the replacement property. If you make the election but do not spend all of your compensation on qualifying replacement property within the replacement period, you must file an amended tax return and refigure your tax liability for the tax year in which you made the election.Your basis in replacement property is its cost minus any gain that you postpone.

Suggested Reference

U.S. Department of Agriculture, Forest Service. 1989. *Forest Owners' Guide to Timber Investments, the Federal Income Tax, and Tax Recordkeeping (Handbook No. 681, GPO No. 001-00-04540-7).* Superintendent of Documents, U.S. Government Printing Office, Washington, DC 20402. 96 pp.

13

FINANCIAL ANALYSIS OF WOODLAND INVESTMENTS[1]

Woodland investments require a long-range commitment of money, land, time, and other resources. Because such resources are limited, you need to identify and evaluate various investment alternatives to determine how these resources can best be used to meet your demands. This process, called financial analysis, is described in this chapter. Keep in mind, however, that a financial analysis offers one input toward deciding which alternative is best. Your decisions also may depend on other factors that are not easily quantified or profit-motivated.

Financial analysis requires that you apply several formulas. Fortunately, hand-held calculators and computer programs—including some spreadsheet and financial analysis programs—have been developed to simplify some of the computations.[2] Also, tables have been developed to provide interest rate multipliers for various interest rates and number of years. Application of these tools will allow you to focus your time and attention on generating realistic data and thoroughly interpreting the analysis results.

Steps involved in a financial analysis are:

1. Identify the issue and specify objectives.
2. Identify investment alternatives.

(For each alternative complete steps 3 through 11)

3. Define costs, benefits, and their timing.
4. Estimate values for costs and benefits.
5. Make a preliminary assessment of uncertainty.
6. Select the appropriate minimum acceptable rate of return (MARR).
7. Develop a cash flow table.
8. Incorporate tax effects.
9. Discount all cash flows.
10. Calculate and interpret the appropriate measure of investment worth.
11. Complete the assessment of uncertainty.
12. Compare financial profitability and sensitivity analyses.
13. Select the best alternative(s)
14. Implement and monitor the best alternative(s).

A professional forester can provide you with ideas and data for accomplishing many of these steps. In the end, however, you alone decide which investments are best for you.

[1]The primary source of information for this chapter is: Rose, D.W., C. R. Blinn, and G. J. Brand. 1988. *A Guide to Forestry Investment Analysis (Research Paper NC-284).* USDA Forest Service, North Central Forest Experiment Station. 23 pp.

[2]For further information on microcomputer financial analysis software, contact Forest Resources Extension, University of Minnesota, 1530 North Cleveland Avenue, St. Paul, MN 55108.

STEP 1: IDENTIFY THE ISSUE AND SPECIFY OBJECTIVES

When identifying the issue and specifying objectives, it may be useful to write a short description of the situation or to describe the outcome in a clear concise statement. Consider:

- What is the issue?
- Who else is involved?
- Are there any conflicts that need to be addressed?
- What has made you consider taking action?
- What should be the result of that action?

It's important to clearly define your objectives. Be sure to consider:

- Time horizon.
- Need and timing for income.
- Minimum acceptable financial return (expressed as an interest rate).
- Amount of risk you are willing to accept.
- Limits on availability and use of resources (money, land, equipment, staff, etc.).

STEP 2: IDENTIFY INVESTMENT ALTERNATIVES

Identifying all possible courses of action is one of the most important steps in a financial analysis. Begin by developing a list of potential solutions to the issue identified earlier.

The number of alternatives you'll want to evaluate depends on your available time and analysis capabilities. While hand calculations are time-consuming, automated procedures may permit you to analyze a large number of alternatives.

To reduce the number of alternatives to a manageable few, start by considering your resources:

- Technical—equipment, species, etc.
- Economic—relationship between benefits and costs (timing, amount, etc.).
- Commercial—availability of the necessary funds and markets.
- Financial—working capital and other financial obligations.
- Managerial—adequacy of staffing (number, capabilities, etc.).
- Organizational—administrative structure (autonomy, flexibility, etc.).
- Legal/Ethical—relationship with accepted standards and expectations.

You probably lack data to predict outcomes for many alternatives. Many otherwise promising options may be discarded because there is little or no information regarding benefits.

After developing a list of alternatives you wish to seriously consider, go through Steps 3 through 11 **for each alternative independently** to gather information you will need to compare them. Choose among the alternatives in Steps 12 and 13.

STEP 3: DEFINE COSTS, BENEFITS, AND THEIR TIMING

Investments require costs and produce benefits or revenues at various points during the life of the project. Each of these costs and benefits is called an **activity**. For each alternative, you need to identify these activities and the year(s) when they occur.

To obtain the most useful evaluation, define and maintain each activity individually throughout the analysis. This allows a more in-depth assessment of an investment alternative and permits you to evaluate individual costs and benefits to ensure that the use of resources is economically efficient. It also allows you to focus your attention on particular critical investment activities once you've begun a project. However, for tax planning purposes, you may want to lump net cash flows in some cases (e.g., for each harvest, subtract timber sale preparation and administration costs from revenues to create one activity). You also might lump cash flows where a certain set of activities is identical for all investment alternatives (e.g., a site has to be prepared and planted in one common manner).

Do not include the cash flow from an existing or completed project as an initial benefit in an analysis. However, that income may be used as available funds to cover investment costs.

As you define the time of activities, you must assume that all costs and benefits occur either at the **beginning** or **end** of the year. The first (initial or base) year for an alternative in which all activities are assumed to occur at the end of the year (Convention 1 in Table 13) is "0." Year 0 is the point to which all future cash flows are adjusted in Step 9 of the financial analysis. For analyses where all activities are assumed to occur at the beginning of the year (Convention 2), the initial investment period is designated as "1" and all cash flows are adjusted back to this point.

Table 13. Comparison of the two conventions for indicating occurrence of cash flow activities.

CONVENTION	TIME OF ACTIVITY	YEAR OF PROJECT					
1	END OF YEAR	0	1	2	3	4	5
2	BEGINNING OF YEAR	1	2	3	4	5	6

For example, if you plant Christmas trees and allow them to grow for 10 years and then harvest them, under Convention 1 you plant in Year 0 and harvest in Year 10; under Convention 2 you plant in Year 1 and harvest in Year 11. In both cases 10 years elapse between planting and harvest. When applied consistently throughout a financial analysis, the two conventions lead to identical results. **Examples in this chapter assume all cash flows occur at the end of the year (Convention 1).**

STEP 4: ESTIMATE VALUES FOR COSTS AND BENEFITS

After you have identified the activities and year(s) of occurrence, you need to attach monetary values to each activity. Both the number of units (e.g., volume removed) and the per-unit cash flow must be estimated for each occurrence of an activity. You will need reasonable estimates to minimize the uncertainty associated with your financial analysis. While prices for some products may be available through published reports, a knowledge of local markets and conditions is useful.

When defining your cash flow values, make sure that they all are expressed in the same units within each investment alternative (e.g., $/acre).

Regardless of when an activity occurs, assign it a cash flow that is the estimated value of that activity **when the project begins.** That is, estimate values in "today's (or base year) dollars." Use cash flows that represent the number of units multiplied by the cash flow per unit. As an example, assume that an income that is to occur in Year 10 will yield fifty units. The current value of each unit is $20. Therefore, the value of that activity in today's dollars is $1,000.

STEP 5: MAKE A PRELIMINARY ASSESSMENT OF UNCERTAINTY

Many investors and analysts perform financial evaluations without fully considering the effect of unanticipated variations in cash flows or other assumptions. However, the accuracy of cash flow estimates is always suspect because future developments cannot be completely anticipated. In general, the greater the expected return from an investment, the more you risk losing your money.

The major sources of uncertainty in a woodland financial analysis are:

- Lack of information about future political decisions, resources, and natural processes (e.g., changes in tax laws, zoning, or management practices; cost of labor and materials; growth rate of the timber stand).
- Unpredictable changes in demand for products (e.g., supply and demand shifts for individual products; changing local, regional, national, and international markets).
- Catastrophic natural events that can damage resources and alter the natural environment (e.g., drought, disease, high wind, fire).

The cash flow values assumed in an analysis frequently represent potential average values or **expected** outcomes. Actual results may be higher or lower. In most cases, cash flow estimates for activities that occur at the beginning of an investment should be fairly accurate. However, cash flow values become less certain as the time horizon is extended.

Assess the uncertainty in your cash flow assumptions both before (Step 5) and after (Step 11) analyzing each alternative to avoid biasing the investment decision. Perform a preliminary assessment of uncertainty for each cash flow by recording the variability around the assumed value using one of the following techniques:

- Indicate a range of possible percentages (e.g., plus or minus five percent of the assumed value).
- Indicate a range of possible cash flow values (e.g., plus or minus $10/acre of the assumed value).
- Use codes or categories (e.g., C - certain, FC - fairly certain, UC - uncertain, and HU - highly uncertain).

For each activity, provide written comments on what factors cause the uncertainty. Don't be too concerned with estimating the variability associated with activities:

- That occur when the investment begins.
- That have a small cash flow relative to other cash flows.
- For which you can obtain a contract with a fixed value agreed upon in advance.

To apply these principles of estimating cash flows and their relative certainty (Steps 4 and 5), let's assume you are considering planting 20 acres of pine and want to compare this investment to other alternatives. According to your best estimates, costs will include $110 per acre in today's dollars for site preparation and planting and $6 per acre for annual property taxes throughout the rotation. A "thin" activity at the end of the 40th year will yield approximately $800 per acre in today's dollars; the "final harvest" activity at the end of the 60th year has an estimated value of $3,000 per acre today. These cash flows in today's dollars before taxes, as well as their relative degree of uncertainty, are shown in Table 14.

Table 14. Cash flow estimates and the associated degree of uncertainty for a pine plantation.[1]

ACTIVITY	YEAR(S) OF OCCURRENCE	CASH FLOW ($/ACRE)	DEGREE OF UNCERTAINTY[2]
SITE PREP AND PLANTING	0	-110	C
PROPERTY TAX	0 - 60	-6	FC
THIN	40	800	UC
FINAL HARVEST	60	3,000	HU

[1]Cash flow in today's dollars, before tax. Costs are indicated with a minus sign (-).

[2]C - Certain, FC - Fairly certain, UC - Uncertain, HU - Highly uncertain.

Since the "thin" and "final harvest" activities occur many years in the future, there is uncertainty regarding their cash flow values. While the "property tax" activity occurs throughout the life of the alternative, it is a relatively small cash flow.

STEP 6: SELECT THE APPROPRIATE MARR

To be economically worthwhile, an investment must provide a certain rate of financial return, expressed as an interest rate. This rate, which represents the cost of keeping money tied up in an investment, is known as the **minimum acceptable rate of return (MARR)**. If you borrow funds to invest in a woodland management activity, the MARR should be at least equal to the interest rate charged by the lender. If you divert money from your personal savings to invest in woodland management, the MARR is the return you could get by investing your funds in the best alternative investment. As the term suggests, the MARR becomes a guide against which you will be able to accept or reject investment alternatives. Your MARR will be lower if you use personal savings than if you borrow money.

The MARR is calculated as:

$$r = (1 - P)r_b + (1 - B)r_p$$

Where: r = the stated MARR, expressed as a decimal

P = the percent of funds provided from personal savings, expressed as a decimal

r_b = the interest rate on borrowed funds, expressed as a decimal

B = the percent of funds that are borrowed, expressed as a decimal

r_p = the interest rate on the best possible use of person savings, expressed as a decimal

Be sure to select an MARR that you believe is realistic over the life of the project. The MARR must be defined over the same period length as the cash flows were expressed. That is, if cash flows are assumed to occur on an annual basis, the MARR must also be expressed as an annual rate. Also, the same MARR must be applied to each alternative.

As an example, if you could earn 10 percent from a savings account and needed to borrow 70 percent of the funds at 12 percent, your MARR would equal:

$$r = (1 - 0.3)\,0.12 + (1 - 0.7)\,0.10 = 0.114 \text{ or } 11.4 \text{ percent.}$$

Because most financial institutions add interest to your account more than once a year through a process called **compounding**, you probably will need to adjust your MARR to reflect the true rate of return that you are receiving. This true MARR, which is sometimes called an **effective interest rate** or the **annual percentage rate (APR)**, is calculated using:

$$i = [(1 + r/m)^m] - 1$$

Where: i = effective MARR, expressed as a decimal

r = the stated MARR, expressed as a decimal

m = number of compounding periods per year

As an example, assume that the stated MARR (r) is 9 percent and that interest is compounded four times per year ($m = 4$). The effective MARR is:

$$i = [(1 + 0.09/4)^4] - 1 = 0.0931 \text{ or } 9.31 \text{ percent.}$$

In this example, you would use 9.31 percent as the MARR in your analysis.

Financial institutions quote interest rates that include a component for the general inflation rate within the overall economy. Therefore, your MARR also has an inflation rate built into it. You need to estimate what that inflation rate might be over the lifetime of your analysis.

If you are comparing two or more investment alternatives on an after-tax basis, convert the MARR to an after-tax rate. For example, if you are comparing the pine plantation alternative (Table 14) to a bank certificate of deposit that pays 10 percent per year, that 10 percent offered by the bank is your return before taxes. If your marginal tax rate is 28 percent, then your after-tax MARR is 28 percent less, or 7.2 percent. This is the MARR you would use for the rest of the financial analysis.

STEP 7: DEVELOP A CASH FLOW TABLE

Next, develop a cash flow table for each alternative to show cash flows by year of occurrence. These tables serve as a basis for calculating tax effects and the profitability of each alternative. Some computer financial analysis software will produce cash flow tables.

While your original cash flow estimates for each activity were developed in base year dollars, the values that you enter in the cash flow table must be the actual cash flow that occurs in the corresponding year. Therefore, the actual cash flow that occurs in any year following the initial base year must be increased by the general inflation rate that you determined in Step 6 before it is entered into your cash flow table. Use the following formula to inflate a cash flow value in today's dollars to a future value:

$$V_n = V_0(1 + GI)^n$$

Where: V_n = value in Year n (i.e., after interest has compounded for n years)

V_0 = value in Year 0 - the beginning of the investment (present value)

GI = general inflation rate per year within the overall economy, expressed as a decimal

n = number of years over which interest is calculated

Your marginal tax rate does not affect the rate of inflation; therefore, use the same assumed inflation rate for both before-tax and after-tax cash flow tables.

Returning to the pine plantation example (Table 14), let's assume an annual inflation rate of 5.5 percent. If the estimated value of the "thin" activity at the end of Year 40 is $800 in today's dollars, the future inflated value of that activity is calculated as follows:

$$\$800\ (1 + 0.055)^{40} = \$6,810.63$$

The inflated cash flow values for all activities in the pine plantation example are shown in Table 15. The corresponding cash flow table for this example is shown in Table 16. The net benefits row at the bottom of the table summarizes the difference between all of the cash flows in each year.

Table 15. Cash flows inflated to the year of occurrence for the pine plantation example, before tax.

ACTIVITY	YEAR	CASH FLOW ($/ACRE)
SITE PREPARATION AND PLANTING	0	-110.00
PROPERTY TAX	0 - 60	-6.00[1]
THIN	40	6,810.63
FINAL HARVEST	60	74,518.95

[1]The entire cash flow for the property tax activity is not shown because it would require a different value for each of 61 years.

Table 16. Cash flow table for the pine plantation example, before tax.[1]

	ACTIVITY	$/ACRE INFLATED TO YEAR OF OCCURRENCE					
		0	1	2	3	40	60
COSTS	SITE PREP AND PLANTING	110.00	0.00	0.00	0.00	0.00	0.00
COSTS	PROPERTY TAX	6.00	6.33	6.68	7.05	51.08	149.04
	TOTAL	116.00	6.33	6.68	7.05	51.08	149.04
BENEFITS	THIN	0.00	0.00	0.00	0.00	6,810.63	0.00
BENEFITS	FINAL HARVEST	0.00	0.00	0.00	0.00	0.00	74,518.95
	TOTAL	0.00	0.00	0.00	0.00	6,810.63	74,518.95
	NET BENEFITS	-116.00	-6.33	-6.68	-7.05	6,759.55	74,369.91

[1]Only a portion of the cash flow table is shown. The property tax activity occurs annually, inflating at the rate of 5.5 percent per year.

STEP 8: INCORPORATE TAX EFFECTS

When you engage in activities for profit, resulting income and expenses are subject to federal and (in most cases) state income tax regulations. Income taxes will affect the amount and timing of cash flows as well as the MARR, and may influence the profitability of the various alternatives differently, so it is important to perform an after-tax analysis (i.e., to incorporate the effects of taxes) for each alternative. This section will describe how to modify cash flow tables to consider the effects of federal income taxes.

The federal income tax you pay depends on the amount of income you earn and your filing status. Income is divided into brackets and a different tax rate applies to income in each bracket. Tax rates for individuals at this writing are 15 percent, 28 percent, and 33 percent depending on your income and filing status (married filing jointly, married filing separately, single, and head of household).

To analyze an investment on an after-tax basis, reduce all income in the cash flow table by a percentage equal to your marginal tax rate. The formula for calculating an after-tax value is:

$V_{at} = V_{bt}\,(1\text{-}mt)$

Where: V_{at} = value after tax

V_{bt} = value before tax

mt = marginal tax rate, expressed as a decimal

Let's assume your potential income from the pine plantation described in Table 15 would be taxable at a 28 percent rate. On an after-tax basis, your income would be 28 percent lower. You would multiply each income value by 0.72 (i.e., 1.00 - 0.28). After-tax income from the "thin" activity is $4,903.65 and after-tax income from the "final harvest" activity is $53,653.64 (Table 17).

In addition to reducing income in the cash flow table when analyzing an investment on an after-tax basis, you also must reduce expenses by a percentage equal to your marginal tax rate. Because taxes are paid only on net income, a deductible expense lowers the tax payable by a percentage equal to the marginal tax rate multiplied times the expense.

Table 17. Cash flows inflated to the year of occurrence for the pine plantation example, after tax.[1]

ACTIVITY	YEAR(S)	CASH FLOW ($/ACRE)
SITE PREP AND PLANTING	0	-110.00
TAX SAVINGS due to amortization of site prep and planting	0	2.09
TAX SAVINGS due to capitalizing 5% of site prep and planting	1 - 6	4.18
	7	2.09
	60	1.54
10% TAX CREDIT FOR SITE PREP AND PLANTING	0	11.00
PROPERTY TAX	0 - 60	-4.32[2]
THIN	40	4,903.65
FINAL HARVEST	60	53,653.64

[1]Assumes a 28 percent tax rate.

[2]The entire cash flow for the property tax activity is not shown because it would require a different value for each of 61 years.

As you learned in Chapter 12, some expenses are deductible in the year incurred, but others must be capitalized and recovered when the property is disposed of, or are recovered by depreciation or amortization. In an after-tax cash flow table always show the expense in the year incurred; however, the tax savings that results from deducting that expense may not appear until later, especially if the expense is capitalized.

For an active business, operating expenses are fully deductible in the year incurred. That means you should reduce costs for operating expenses that appear in the cash flow table by a percentage equal to the marginal tax rate.

For example, if we assume the property tax of $6 per acre in the pine plantation example is an operating expense for an active business, you can calculate its after-tax value by reducing this cash flow by the assumed marginal tax rate of 28 percent. In Table 17 the property tax in Year 0 after taxes is $4.32 per acre ($6.00 x 0.72). This deduction results in a tax savings of $1.68. If this were a passive business rather than an active business, you would show the full cost ($6.00 in Year 0) in the year it occurred and the tax savings ($1.68 for Year 0) in the year in which you had sufficient passive income to claim the deduction. If this were an investment, you could deduct this expense (i.e., show the tax savings) only to the extent that when aggregated with all other itemized expenses, it exceeded 2 percent of adjusted gross income.

From Chapter 12 we know that tree-planting costs must be capitalized. They cannot be deducted fully in the year trees are planted. Under certain circumstances these costs can be amortized and deducted over 8 tax years and a 10 percent investment tax credit can be claimed.

If we assume the $110 cost for site preparation and planting in the pine plantation example is eligible for amortization and the investment tax credit, then the after-tax cash flow table would show a $110 expense and an $11 tax credit (10 percent of the $110 planting cost) in Year 0.

When claiming a 10 percent investment tax credit, you can amortize 95 percent ($104.50) of the tree planting cost ($110.00); the remaining 5 percent ($5.50) must be capitalized and deducted when trees are sold in Year 60. The amortization schedule for tree-planting costs permits deduction of 1/14 of the amortizable basis ($104.50) in Year 0 ($7.46), 1/7 ($14.93) in Years 1 through 6, and 1/14 ($7.46) in Year 7. Each deduction results in a tax savings equal to our marginal tax rate times the amount of the deduction. In Years 0 and 7 a deduction of $7.46 results in a tax savings of $2.09 (i.e., $7.46 x 0.28). In Years 1 through 6 the deduction of $14.93 results in a tax savings of $4.18 (i.e., $14.93 x 0.28). Do not inflate tax savings that result from amortization and that occur in future years. Tax savings due to the amortization deduction are shown in Table 17.

If we assume you plan to take a 10 percent investment tax credit for tree-planting expenses, then only 95 percent of the tree planting expenses can be amortized, and the remaining 5 percent must be capitalized and recovered when the trees are sold. In the pine plantation example you must carry forward 5 percent of $110 ($5.50) to Year 60. The tax savings that results from that deduction is $1.54 (i.e., $5.50 x 0.28) (Table 17).

STEP 9: DISCOUNT ALL CASH FLOWS

Before you can compare your alternatives, you must convert the cash flows into comparable values. This is done through a process called **discounting**, which is just the opposite of compounding; it starts with a future value, obtained from your cash flow table, and finds its worth today. The comparable values, known as **measures of investment worth**, can be used as criteria against which to measure alternative courses of action.

Use MARR as the discount rate to convert future cash flows to their present value equivalent. Before performing any discounting calculations, first sum the bottom row of each cash flow table (the net benefits row). (If you are performing an after-tax analysis, only do this for your after-tax cash flow table.) A negative total indicates that the alternative cannot even provide a 0 percent rate of return, thereby eliminating it from further consideration. Do not bother to perform Steps 9 through 11 for those alternatives.

The basic formula for discounting is:

$$V_0 = V_n/(1 + i)^n$$

Where: V_0 = value in Year 0 - the beginning of the investment (present value)

V_n = value in Year n (i.e., after interest has compounded for n years)

n = number of years over which interest is calculated

i = interest rate (MARR)

As an example, the after-tax present value at a MARR of 7.2 percent for a thinning activity which has a value of $4,903.65 (Table 17) in Year 40, is $303.90:

$$V_0 = \$4,903.65/(1.072)^{40} = \$303.90$$

Thus, $4,903.65 received 40 years in the future is worth $303.90 today. You do not need to discount cash flows that occur in the base year.

After discounting all of the values in your cash flow table, sum the present values of your discounted expenses. Do the same for incomes. You will use the resulting values to calculate your measure of investment worth.

STEP 10: CALCULATE AND INTERPRET THE APPROPRIATE MEASURE OF INVESTMENT WORTH

Measures of investment worth help us assess the profitability of each investment alternative. The most commonly used measures of financial worth for forestry investments are net present value (NPV), internal rate of return (IRR), equivalent annual income (EAI), and soil expectation value (SEV).

In any given situation, there is one measure of investment worth that is most appropriate. Determine which measure is most appropriate for your decision and then calculate only that measure of investment worth for each alternative.

Net Present Value

Net present value is the discounted value of all benefits minus the discounted value of all costs. NPV uses the MARR to discount all costs and benefits back to the base year. NPV is calculated as:

$$NPV = \sum \frac{B_n}{(1+i)^n} - \sum \frac{C_n}{(1+i)^n}$$

Where: $\sum$ = sum of all values over all years in the analysis

B_n = benefit activity cash flow in Year n

C_n = cost activity cash flow in Year n

i = interest rate (MARR)

Preferred alternatives are those that have a positive NPV, since they will yield a higher return than the MARR. The value of the positive NPV represents the amount in base year dollars that the investment will return beyond the MARR. In choosing among alternatives, you generally would select the one with the highest NPV, depending upon the sensitivity analysis results (see Step 11 below) and your personal preferences.

Let's assume you are considering the pine plantation example (Table 17) and want to know whether this investment is better than an alternative that will yield a 7.2 percent rate of return (MARR) after tax, where the assumed rate of inflation is 5.5 percent per year. A financial analysis computer program calculates the NPV to be $886.18 per acre after tax. The positive NPV for this alternative indicates that it meets the after-tax MARR of 7.2 percent. In addition, it would also provide $886.18 per acre after taxes in today's dollars.

Internal Rate of Return

The IRR is the rate of return for which the present value of benefits just equals the present value of costs. It is the interest rate at which NPV is zero. Unless you have access to an automated tool for calculating IRR, it is best solved through a trial and error process. The formula for calculating IRR is:

$$IRR = \sum \frac{B_n - C_n}{(1+i)^n} = 0$$

Alternatives with an IRR greater than the MARR are preferred. In choosing between two alternatives, select the one with the higher IRR if the sensitivity analyses are nearly equivalent.

For the pine plantation investment (Table 17), you know that NPV at 7.2 percent is $886.18 after taxes. Because the NPV is positive at that interest rate, IRR must be greater than 7.2 percent. At a 12 percent after-tax interest rate (still assuming a 5.5 percent general inflation rate), NPV is -$38.80 per acre, indicating that the IRR must be less than 12 percent. If you try an after-tax rate of 11 percent, NPV equals $16.35. Therefore, the IRR must be slightly greater than 11 percent. Through trial and error you will find the after-tax IRR is 11.24 percent (assuming a 5.5 percent general inflation rate). This means that the plantation investment will yield a positive NPV for all after-tax MARRs less than 11.24 percent. If your after-tax MARR is less than or equal to 11.24 percent, you would consider implementing this investment alternative.

Equivalent Annual Income

An annual income or loss is sometimes a more useful measure of investment worth than the lump sum value at the beginning of the investment period, as calculated by NPV. The equivalent annual income is the net present value converted to an annual value paid at the end of each year over the life of the investment. EAI is useful for comparing alternatives that yield a periodic return (e.g., timber) with those that yield an annual return (e.g., agricultural crops) **provided that the lives of the alternatives are equal.**

Before you determine the EAI, you first calculate the interest rate that represents the difference between your MARR and the general inflation rate (GI). This interest rate, which is sometimes known as the real rate (RR) is calculated as:

$$RR = \left(\frac{1+MARR}{1+GI}\right) - 1$$

For the pine plantation example (Table 17), the after-tax real rate equals:

$$RR = \left(\frac{1+0.072}{1+0.055}\right) - 1 = 0.0161 \text{ or } 1.61\ \%$$

To calculate EAI in base-year dollars, use the following formula:

$$EAI = NPV\left(\frac{RR\,(1+RR)^{n}}{(1+RR)^{n}-1}\right)$$

Returning to the pine plantation example, the NPV is $886.18 at a 7.2 percent MARR after-tax. EAI is then calculated as:

$$EAI = \$886.18\left(\frac{0.0161\,(1+0.0161)^{60}}{(1+0.0161)^{60}-1}\right) = \$23.14$$

The pine plantation will earn a 7.2 percent MARR after-tax (assuming a 5.5 percent general rate of inflation) plus $23.14 per acre per year in base year after-tax dollars at the end of each year over the course of the 60-year investment.

Soil Expectation Value

Soil expectation value is the present value of an infinitely long series of cash flows resulting from land management. It represents the maximum amount that you could pay in base year dollars for a tract and still earn the MARR. It is useful for estimating how much you should bid for bare land that would be used for growing successive crops and for comparing management alternatives that have different rotation lengths. SEV assumptions are:

- All costs within one cycle are included in the analysis, including relevant management fees, administrative costs, and taxes
- The land will be managed for the same land use option forever and the same prescription will be used for each subsequent production cycle
- The rotation starts with bare land
- The values of all costs and benefits are identical for all rotations, in base year dollars
- Land value does not enter into the calculation

SEV in base year dollars is calculated using the following formula:

$$SEV = NPV + \left(\frac{NPV}{(1+RR)^{n}-1}\right)$$

For the pine plantation example (Table 17), we will assume that it is appropriate to calculate SEV because the alternative starts with bare land. The SEV is calculated as:

$$SEV = \$886.18 + \left(\frac{\$886.16}{(1+0.0161)^{60}-1}\right) = \$1{,}437.54$$

If this investment alternative were repeated in the exact same manner for an infinite number of rotations, you could pay $1,437.54 per acre to purchase the land today and still earn a 7.2 percent after-tax rate of return. If the land could be purchased for less than $1,437.54 per acre, you would likely obtain a greater rate of return.

Determine Which Measure to Use

There is some disagreement about the relative merits and applications of the measures of investment worth; information presented here represents our recommendations. Two factors that you'll need to consider in deciding which measure you will use are independence and relative length of the alternatives.

Independence of Alternatives

Investment alternatives can be either mutually exclusive or independent.

Mutually Exclusive Alternatives

Alternatives are mutually exclusive if, for a given location, the choice of one alternative excludes all others. NPV is the preferred measure of investment worth for choosing among mutually exclusive alternatives, except where the length of the alternatives are not equal, as discussed in the next section. Either EAI or SEV may be used instead. If the alternatives have similar risks, you would prefer the one with the largest positive NPV.

Scale differences become a factor if you could accomplish a task by spending a little money or a lot more money (e.g., spend $20 per acre or $200 per acre to accomplish the same task). If there is a difference in scale between alternatives, you need to determine the scale that enables you to capture the maximum NPV (or EAI) while still meeting your MARR. To do this, array the alternatives by scale from smallest to largest, calculate the NPV or EAI associated with each, and then determine the incremental change in NPV or EAI that occurs by going from one scale to the next higher. The most financially efficient scale is that one with the smallest positive NPV or EAI scale advantage. This is the scale at which you capture all potential NPV or EAI while still meeting your MARR for the incremental investment. The closer the difference in NPV or EAI advantage is to zero, the closer you will come to maximizing your returns for the selected scale; you may not be able to achieve a zero NPV or EAI advantage if it is not possible to produce small increments of change in scale.

As an example of evaluating scale differences, assume you want to choose among three pond sizes, that your MARR is 11 percent (assuming a 5.75 percent general inflation rate), and that the life of the investment is 15 years (Table 18). NPV is used as the decision criterion.

Table 18. Analysis of scale differences for pond sizes.

POND SIZE	INITIAL COST	INCREMENTAL INVESTMENT	ANNUAL BENEFIT	NPV	NPV advantage of the additional investment
	DOLLARS				
Small	500	0	100	640.46	0
Medium	1,000	500	170	938.78	298.32
Large	1,500	500	210	894.96	- 43.82

In this example, going from the small to medium scale earns you an additional $298.32 in NPV. However, when building the large pond, the additional costs fail to be recovered with an 11 percent nominal MARR by $43.82. There is some pond size between medium and large that will maximize returns (the NPV advantage of the additional investment would be $0 for that pond size). This optimal pond is closer to the large size than to the medium size.

Independent Projects

Independent projects are investments where two or more alternatives can be chosen simultaneously; each alternative is evaluated on its own merit and may be accepted if it meets your criteria, regardless of whether other alternatives also are accepted. Comparisons of independent investment proposals identify which alternatives satisfy some minimum set of qualifications; all such alternatives could be implemented. Individuals or organizations that manage multiple parcels of land are most likely to evaluate independent projects. Building a recreational trail, buying a tree-planting machine, and constructing a storage shed for fire control tools are examples of independent projects.

For the case of independent projects where capital is **not** limiting, any criterion can be used for selection.

For the case of independent projects where capital is limiting **(capital budgeting)**, allocate capital based upon how you intend to fund your alternatives; that is, whether you intend to borrow money or only plan to use personal savings. If you plan to borrow money, your MARR would be the borrowing rate of interest and you would tentatively select all projects with a non-negative NPV or EAI. From that tentative list of projects, add each project's budget to get the tentative total investment. If that tentative total exceeds the savings that you want to invest in projects, borrow an amount equal to the budget deficit at your MARR and fund the entire group of projects tentatively selected.

If you intend to only use personal savings, your MARR would be the best rate that you could get from your best alternative investment. Once again, you would tentatively select all projects with a non-negative NPV or EAI and develop a tentative total budget. If this tentative total budget is less than your available personal savings, fund the entire tentative list of projects and invest surplus funds elsewhere. However, if the total tentative investment budget is greater than your personal savings, order the projects by decreasing IRR and fund the most acceptable projects until your personal savings are spent. If you do not fully expend your savings and it is possible to implement a portion of the next project that would cause you to exceed your investment budget, select that partial project and implement the appropriate portion. Otherwise, invest surplus funds elsewhere.

Relative Length of the Alternatives

You cannot directly use NPV to compare investment alternatives that do not have the same investment life (number of years in the analysis). The reason for this is obvious if you consider the analogy of the choice between paying $10 for a one-year fishing license or paying $25 for a three-year license; a simple comparison of $10 and $25 is inaccurate since the extra $15 buys two more years of fishing.

If all assumptions are met, SEV is an appropriate measure for comparing unequal-life investment alternatives since it allows an automatic comparison of projects of any length. Fortunately, the increasing uncertainty over time is made less important because early cost and revenues contribute most to SEV. Use of higher MARRs also lessens the impact of future uncertainty on SEV.

EAI can be used to compare alternatives that have different lengths if you can assume that projects with a shorter investment horizon can be continued in the future with the same cash flow and interest rate assumptions.

Deciding Which Alternative(s) Should be Further Evaluated

After deciding which measure is most appropriate to use, you need to decide which alternatives you will further evaluate. The formal decision rule is to further consider those investment alternatives that have a non-negative NPV, EAI, or SEV or have an IRR greater than or equal to the MARR. Remember, though, that because of the uncertainty associated with a financial analysis, alternatives that do not appear to be worth considering could be financially attractive with a different set of assumptions. Therefore, you may want to further evaluate alternatives that do not initially appear to be worthy of further consideration if you have sufficient time, if you have microcomputer software available to perform the sensitivity analysis, if they are not too financially unattractive, or if you are still interested in them.

STEP 11: COMPLETE THE ASSESSMENT OF UNCERTAINTY

In a financial analysis, the critical point for any investment alternative is the point at which changes in assumptions in cash flows and MARR cause the proposal to change from acceptable to unacceptable. Therefore, you should apply a **sensitivity analysis**—an examination of several different cash flow and interest rate scenarios—to investment alternatives that meet your MARR and to those potentially unattractive alternatives that you decided to further evaluate. The objective of this analysis is to determine how much your measure of investment worth changes when you change your assumptions. Use after-tax assumptions for the sensitivity analysis if you are incorporating taxes into your financial analysis.

Cash Flows

A sensitivity analysis for cash flows shows the effect of a specified change in today's dollars on your measure of investment worth for each investment activity. To test the sensitivity of your measure of investment worth to changes in cash flow assumptions, vary the cash flow for one activity and then recalculate your measure of investment worth as follows:

1. Change the initial cash flow estimate for one activity by 10 percent.
2. Recalculate your measure of investment worth and record the new value.
3. Calculate and record the absolute amount of change that results from this analysis (e.g., difference in NPVs). If a cost increases or a benefit decreases, your measure of investment worth will be less favorable by this amount. If a cost decreases or a benefit increases, your measure of investment worth will be more favorable by this amount.
4. Reset the cash flow to the original value.
5. Return to Step 1 above, changing another cash flow. Perform Steps 1 through 3 until you have individually evaluated the sensitivity of all activities that you wish to test.

Finally, compare the results of this analysis to your preliminary assessment of uncertainty. If the analysis indicates that the alternative is very sensitive to realistic changes in cash flow assumptions, you might choose not to select that alternative because it has a high amount of uncertainty associated with it. Or, you might choose to collect more data about that alternative before making a final decision. If you do select the alternative, you should closely monitor investment performance after implementation.

While the sensitivity analysis may indicate that a particular cash flow is especially sensitive to changes in assumptions, you do not need to be too concerned about activities that occur when the investment is implemented, that have a small cash flow relative to other cash flows, or for which you can obtain a written contract with a fixed cost or revenue agreed upon in advance.

To minimize the number of calculations necessary for your sensitivity analysis, begin by evaluating activities that your preliminary analysis determined were uncertain or highly uncertain, that do not occur near the beginning of the alternative, and/or that have a relatively large cash flow associated with them. If any of these activities cause you to become so concerned about the viability of the alternative that you would not implement it, there is no need to perform further calculations for that alternative.

For the pine plantation example (Table 17), Table 19 shows that our measure of investment worth, NPV, is most sensitive to a change in the cash flow for the "final harvest" activity. A 10 percent decrease in revenue from the "final harvest" activity would reduce NPV by $82.78 per acre in today's dollars after taxes. Conversely, a 10 percent increase in revenue would increase NPV by $82.78 per acre in today's dollars after taxes. Since the "final harvest" activity occurs 60 years in the future, there is a high degree of uncertainty about the amount of revenue it will generate (Table 14).

Table 19. NPV sensitivity to a 10 percent change in cash flow assumptions for the pine plantation example. Values are shown in today's dollars after taxes.

ACTIVITY	NPV CHANGE ($/ACRE)
SITE PREP AND PLANTING[1]	7.58
PROPERTY TAX	16.97
THIN	30.39
FINAL HARVEST	82.78

[1]Reflects the net effects of combining the cost for site preparation and planting with the tax savings from amortizing 95 percent of the cost, capitalizing 5 percent of the cost, and claiming a 10 percent investment tax credit.

The amount and percent of change necessary in today's dollars for each cost and revenue activity to force NPV to 0 also can be calculated using the following formula:

C = (V/A) P

Where: C = percent change from the original cash flow estimate required to just return the MARR

V = original value for your measure of investment worth ($/acre)

A = amount of change in your measure of investment worth resulting from the specified change ($/acre)

P = percent change specified in the sensitivity analysis

For the "final harvest" activity in the pine plantation example, the percent change from the original cash flow estimate required to force NPV to 0 equals:

C = ($886.18/$82.78) (10) = 107.05 percent

Table 20 shows how much the value of each activity in the pine plantation example would have to change to force NPV to 0 (up to a maximum of 100 percent).

Table 20. After-tax percent change in individual activity cash flows, up to a maximum of 100 percent, required to make NPV equal zero for the pine plantation example. Value changes are shown in today's dollars.

ACTIVITY	PERCENT CHANGE[1]	DOLLAR CHANGE ($/ACRE)
SITE PREP AND PLANTING	100.00	- 75.80
PROPERTY TAX	100.00	- 169.67
THIN	-100.00	- 303.90
FINAL HARVEST	-100.00	- 827.77

[1]When percent change equals ±100 percent and the dollar change is less than $886.18, the activity will not alter the overall investment selection, using NPV as the measure of investment worth.

On an after-tax basis, more than a 100 percent change in any one cost or revenue activity would be required in order for NPV to drop to 0 for the pine plantation example. The final decision to invest will depend on how confident you feel about individual costs and benefits and how much risk you are willing to take.

Interest Rates

To assess uncertainty in interest rates, recalculate your measure of investment worth using different assumptions for MARR and (in an after-tax analysis) for the marginal tax rate. You do not need to test different assumptions for the general inflation rate.

Start by changing one of your interest rate assumptions and then noting the new value for your measure of investment worth. Hold all cash flow assumptions constant. Repeat the process for different feasible interest rate assumptions. Once again, you are concerned about the point at which the investment fails to meet your decision criterion.

When NPV was first calculated for the pine plantation example, the resulting value was $886.18/acre. Table 21 shows different results when the after-tax NPV is recalculated using various MARRs, each of which still assumes a 5.5 percent general inflation rate and a 28 percent marginal tax rate. At an after-tax MARR of 12 percent and a general inflation rate of 5.5 percent, the alternative does not meet the MARR requirement because the NPV is negative. If you felt that this after-tax MARR was realistic, you might not want to invest in this alternative.

Table 21. After-tax NPVs at various MARRs for the pine plantation example where the general inflation rate remains constant at 5.5 percent and the marginal tax rate is held constant at 28 percent.

AFTER-TAX MARR (PERCENT)	AFTER-TAX NPV ($/ACRE)[1]
12.00	- 38.80
11.00	16.35
10.00	109.59
9.00	267.74
8.00	537.33
7.00	999.80

[1]Cash flow in today's dollars.

STEP 12: COMPARE FINANCIAL PROFITABILITY AND SENSITIVITY ANALYSES

After calculating your measures of investment worth and evaluating the sensitivity of your analysis for each alternative, compare the alternatives. The goal here is to rank the alternatives according to these two factors to help you later select the best one(s).

In addition to comparing the financial profitability and the sensitivity analysis results, also consider any other factors that affect your decision. Rank projects according to the various criteria, arraying them from best to worst. Indicate for which alternative(s) you are willing to accept the associated uncertainty.

STEP 13: SELECT THE BEST ALTERNATIVE(S)

Merely completing the steps in a financial analysis will not always produce a final decision. You need to consider all of the factors that affect your decision. While it may be difficult to quantify the factors that are not purely financial in nature, they need to be incorporated before you select the alternative(s) with which you feel most comfortable. While you must be extremely careful to do an accurate analysis, in the long run, it is human judgment that counts the most.

If you are choosing among mutually exclusive alternatives, select the one best alternative. However, if you are selecting among independent alternatives, there may be more than one acceptable alternative. Select the alternative(s) with which you feel most comfortable.

STEP 14: IMPLEMENT AND MONITOR THE BEST ALTERNATIVE(S)

Now that you have selected an alternative, that doesn't mean your work is over! You now must follow the plan that you devised when identifying activities and cash flows for the alternative. Because an alternative is seldom implemented exactly as indicated in the original plan, you may want to develop and maintain an historical log book that contains information discussing what actually happened, who did what, the specific dates when activities occurred, the associated costs or returns, and any explanations. This log book will become more important in future years as tax savings need to be considered. Maintain separate log books for each piece of property that you own or manage.

When you implement an alternative, you have made the decision that this alternative is the best among all that are currently available. However, in the future, as circumstances change, this alternative may no longer be the best, or it might be improved. Therefore, monitor the progress of each implemented alternative to determine whether it should be continued, altered, or terminated.

Before evaluating the performance of your investment(s), update your objectives (see Step 1). Decisions based upon your monitoring must be consistent with your current and future objectives. The monitoring process also requires that you determine the appropriate MARR for that point in time and that you update future cash flow estimates. Estimates of future cash flows are expressed in terms of current-year dollars (i.e., the year in which you are performing the analysis). You should perform a preliminary assessment of uncertainty for all future cash flows.

The **abandonment test** is a useful procedure for monitoring investment performance. This test first requires that you recalculate NPV using updated interest rates and estimates for cash flows that occur today or in the future. (Exclude all past cash flows from the analysis.) The base year is the current year. Because previous cash flows are excluded from the analysis, your alternative frequently will yield a more financially attractive NPV than you may have calculated when first evaluating the project.

Next, calculate the abandonment value for the project (i.e., its current liquidation value) at this time. The abandonment value would be either the value of the liquidated assets or the value associated with investing these assets in another, more profitable investment alternative. The liquidation value might consist of the sale of equipment and machinery, the harvest of all timber, etc.

If the abandonment value is higher than the value for your revised NPV, either abandon the project or alter it to make it more profitable based upon data that became available during the monitoring process.

You may want to perform the abandonment test annually for investments where there are several annual cash flows, where new abandonment values frequently become available, or where future cash flows no longer look realistic. For all other investments, it is still a good idea to perform this analysis every 3 to 5 years.

Application of the abandonment test can be shown for the pine plantation example (Table 14). Assume that the project was implemented, you are currently at the end of Year 40, and you have just completed the "thin" activity. You determine that your current after-tax MARR is 8 percent, including a general inflation rate of 6 percent. You develop a table of actual and estimated future cash flows for the project (Table 22).

While all cost estimates turned out to be the same as originally estimated, the cash flows for both revenue activities have been revised downward due to lower current and projected stumpage prices.

Table 22. Original and actual or revised before-tax cash flow estimates for the pine plantation example after 40 years.[1]

After Tax MARR: 8% **Assumed General Inflation Rate: 6%**

ACTIVITY	YEAR(S) OF OCCURRENCE	CASH FLOW ($/ACRE)	
		ORIGINAL ESTIMATE	ACTUAL OR REVISED ESTIMATE
SITE PREPARATION AND PLANTING	0	- 110	- 110
PROPERTY TAX	0 - 60	-6	-6
THIN	40	800	600
FINAL HARVEST	60	3,000	2,250

[1]All future cash flow estimates are in current year (i.e., Year 40) dollars.

Tax Implications of Abandonment

Income taxes will affect cash flows used in calculating a project's abandonment value. Therefore, use after-tax cash flow tables as the basis for calculating NPV when performing the abandonment test.

Incorporate your marginal tax rate when developing your MARR and cash flow table. Alter the abandonment value that you use to claim as a benefit the tax savings that result from deducting the original cost basis of a capital asset and other capitalized expenses that have not yet been recovered through depreciation or amortization. This may increase the abandonment value.

Assume that the pine plantation example could be abandoned at the end of Year 40 by selling the property for $1,000/acre. Assume that the land was purchased for $124/acre. To decide whether to continue the investment or to abandon it, compound the cash flows for the "final harvest" activity and all future occurrences of the "property tax" activity by 6 percent for 20 years and then discount the result back to the current year (Year 40) using the after-tax MARR of 8 percent. The tax savings from capitalizing 5 percent of the site preparation and planting costs should still be included but not compounded. No other activities should be included in this analysis because their cash flows have already occurred. The after-tax cash flow estimates for the "continue" option are shown in Table 23. The results of this analysis indicate that the revised after-tax NPV for this option is $1,043.61/acre.

Table 23. Pine plantation example after-tax cash flow estimates for the "continue" option of the abandonment test at the end of year 40.[1]

ACTIVITY	YEAR(S)	CASH FLOW ($/ACRE)
PROPERTY TAX[2]	1 - 20	- 4.32
TAX SAVINGS due to capitalizing 5% of site prep and planting[3]	20	1.54
FINAL HARVEST	20	1,620.00

[1]Cash flow in today's (Year 40) dollars, after-tax. A 28 percent tax rate is assumed.

[2]The entire cash flow for the property tax activity is not shown because it would require a different value for each of 20 years.

[3]Capitalized cash flows are not subject to inflation. The value shown is the actual cash flow that would occur in Year 20.

On an after-tax basis, the land sale revenue for the "abandon" option would be $720 ($1,000 x 0.72). You would also earn a tax savings of $34.72 ($124.00 x 0.28) from deducting the original cost basis of the land and a tax savings of $1.54 ($5.50 x 0.28) from deducting the capitalized site preparation cost. The total after-tax value of the "abandon" option is $756.26/acre (Table 24). Because the revised after-tax NPV for the "continue" option is more than the after-tax abandonment value of $756.26/acre, you would make a preliminary recommendation to continue the project.

Table 24. After-tax value of land under the "abandon" option.

ACTIVITY	CASH FLOW ($/ACRE)
LAND SALE	720.00
TAX SAVINGS from deducting original cost basis of land	34.72
TAX SAVINGS from deducting 5% of capitalized site preparation and planting cost	20.16
TOTAL	756.26

Sensitivity Analysis

Your abandonment value generally is known with a relatively high degree of certainty because it is in "today's dollars." However, future cash flow and interest rate assumptions used in updating your NPV will still have some uncertainty.

Therefore, it is still important that you perform a preliminary assessment of uncertainty before recalculating NPV and that you then evaluate the sensitivity of NPV to changes in assumptions. You may find that there is so much uncertainty associated with continuing the project that you would prefer to take a lower abandonment value today.

Closely monitor the performance of an investment that has a relatively large amount of uncertainty but is still financially attractive enough to continue. Perform monitoring analyses annually if you are very concerned about future performance. For the pine plantation example, sensitivity analysis might indicate that you should abandon the project in Year 40 because of uncertainty associated with the "final harvest" activity, particularly since both project revenue cash flows already have decreased due to a reduction in stumpage prices.

Suggested References

Rose, D. W., C. R. Blinn, and G. J. Brand. 1988. *A Guide to Forestry Investment Analysis (Research Paper NC-284).* USDA Forest Service, North Central Forest Experiment Station, 1992 Folwell Ave., St. Paul, MN 55108. 23 pp.

Gunter, J. E., and H. L. Haney, Jr. 1984. *Essentials of Forestry Investment Analysis.* Virginia Polytechnic Institute and State University, Blacksburg, VA 24061. 337 pp.

APPENDIX A
SOURCES OF FORESTRY ASSISTANCE

You should seek assistance from a forester before conducting any management activities in your woodland. Forestry information and assistance are readily available. Described below are the major public agencies, private businesses, and associations that commonly assist private woodland owners in the Upper Midwest.

PUBLIC SOURCES

The **extension service** of the land grant university in each state provides research-based information to woodland owners through conferences, workshops, tours, personal letters, telephone calls, publications, newspaper and magazine articles, exhibits, radio, and television. If you have specific questions about forest management, forest products, or wildlife, contact your county extension office (listed in the telephone directory under your county government offices) or write to the extension forester at the closest university listed below:

University of Illinois
W-503 Turner Hall
1102 South Goodwin
Urbana, IL 61801

Purdue University
1159 Forestry Building
West Lafayette, IN 47907

Iowa State University
251 Bessey Hall
Ames, IA 50011

Kansas State University
2610 Claflin Road
Manhattan, KS 66506

Michigan State University
126 Natural Resources Building
East Lansing, MI 48824

University of Minnesota
1530 North Cleveland Avenue
St. Paul, MN 55108

University of Missouri
I-34 Agriculture Building
Columbia, MO 65211

University of Nebraska East Campus
101 Plant Industry
Lincoln, NE 68583

North Dakota State University
266B, Loftsgard Hall
Fargo, ND 50878

Ohio State University
2021 Coffey Road
Columbus, OH 43210

South Dakota State University
Box 2207-C
Brookings, SD 57007

University of Wisconsin
Russell Labs
1630 Linden Drive
Madison, WI 53706

State foresters usually will examine woodlands, prepare management plans and tree planting recommendations, mark timber for improvement cutting and firewood harvesting, sell seedlings for reforestation, and provide advice on erosion and sediment control, wildlife habitat improvement, insect and disease control, forest recreation, road construction, and urban and community forestry. There generally are no charges for these services; however, state foresters may charge small fees for services pertaining to timber sales in some states and may ask owners to retain a consulting forester for large, commercial timber sales or damage or trespass appraisals.

State forestry offices usually are listed in the telephone directory under state government offices. Below are addresses for state foresters in the Upper Midwest.

Illinois Department of Conservation
Division of Forest Resources
600 North Grand Avenue West
Springfield, IL 62706

Indiana Department of Natural Resources
Division of Forestry
613 State Office Building
Indianapolis, IN 46204

Iowa Department of Natural Resources
Wallace State Office Building
Des Moines, IA 50319

Kansas State University
Department of Forestry
2610 Claflin Road
Manhattan, KS 66502

Michigan Department of Natural Resources
Forest Management Division
Box 30028
Stevens T. Mason Building
Lansing, MI 48909

Minnesota Department of Natural Resources
Division of Forestry
500 Lafayette Road
St. Paul, MN 55155

Missouri Department of Conservation
2901 West Truman Blvd.
P.O. Box 180
Jefferson City, MO 65102

University of Nebraska
Dept. of Forestry, Fisheries, & Wildlife
101 Plant Industry
Lincoln, NE 68583

North Dakota Forest Service
First and Brander
Bottineau, ND 58318

Ohio Department of Natural Resources
Division of Forestry
Fountain Square
Columbus, OH 43224

South Dakota State Forester
445 East Capitol
Pierre, SD 57501

Wisconsin Department of Natural Resources
Bureau of Forestry
101 S. Webster Street
P.O. Box 7921
Madison, WI 53707

Soil and Water Conservation Districts (SWCDs) have slightly different names in each state. They can help you locate natural resource information and programs, technical assistance, and funding for soil conservation and forestry practices. Many SWCDs sell trees and shrubs for erosion control and wildlife habitat. SWCD offices usually are listed in the telephone directory under county government offices.

The **USDA Agricultural Stabilization and Conservation Service (ASCS)** offers cost-sharing funds for a wide variety of soil, water, wildlife, and forestry conservation practices. Technical assistance to plan and design land management practices eligible for cost-sharing are provided by other agencies such as the Soil Conservation Service or your state forester. Contact the local ASCS office found in the telephone directory under U.S. government offices.

The **USDA Farmers Home Administration (FmHA)** offers loans or loan guarantees to establish forestry practices such as thinning, pest control, and fire protection; to purchase farm and forestry equipment, labor, or materials; to buy, improve, or enlarge farms and farm buildings; and to develop outdoor, income-producing recreational facilities. To be eligible, landowners or operators must rely on farm income and any other income needed to provide a living standard comparable to that considered adequate for the area and be unable to obtain reasonable credit terms and rates from private sources. Contact the local FmHA office listed in the telephone directory under U.S. government offices.

The USDA Forest Service, North Central Forest Experiment Station conducts research in forestry and related fields throughout a seven-state area in the north-central United States. Specific areas of study include forest management, silviculture, fire management, economics, forest engineering, wood utilization, insect and disease management, tree physiology and genetics, resource evaluation, wildlife habitat management, and water quality management. The station issues more than 250 new publications annually reporting research results. Publications are free as long as the supply lasts. Write the station at 1992 Folwell Avenue, St. Paul, MN 55108.

The USDA Soil Conservation Service (SCS) provides technical assistance to woodland owners through soil and water conservation districts. Assistance is available for erosion and sediment control practices and soil interpretations that indicate erosion hazard, equipment limitation, expected tree seedling mortality, windthrow hazard, plant competition, site indices of important tree species, and tree species recommended for planting on each soil type. The SCS also assists woodland owners with land-use planning, including assistance with locating access roads, stabilizing grades, building water control structures, reclaiming abandoned mines, and managing wildlife habitat. In addition, it publishes soil surveys that you can use to assess the productivity of your land for timber, wildlife, and agricultural purposes. Some financial aid is available for resource conservation and development projects and for Public Law 566 watershed projects. Contact your local SCS office listed in the telephone directory of your county seat under U.S. government offices.

PRIVATE SOURCES

Forest industries provide the market for timber produced on private woodlands. Some large forest products companies in the paper and waferboard industries and a few large sawmills employ foresters who provide services for woodland owners. Timber management services may include providing forest management plans, harvesting plans, timber tax assistance, reforestation guidance, and information on government cost-share programs.

Some companies have formal landowner-assistance programs. In return for company services, a woodland owner grants the company access to carry out management recommendations developed by the forester and to harvest timber for a specified time. Other companies provide their services under a less formal arrangement. To obtain a list of industrial foresters, contact your state forester or extension service.

Forestry consultants provide a wide range of services including timber appraisals, marketing assistance, management plans, timber stand improvement, tree planting, and pest management. They may be general practitioners or specialists. Good consultants have a college education in forestry and several years of experience, subscribe to a code of ethics developed by their professional organization, and have no business connection with a forest products company where a conflict of interest might occur. Fees are based on an hourly rate, a percentage, a piecework rate, or other mutually agreed upon basis. This fee is a management expense that you usually can deduct from timber income, damage award, or other taxable income.

For a list of forestry consultants, contact the National Association of Consulting Foresters at 5410 Grosvenor Lane, Suite 120, Bethesda, MD 20814. State foresters or extension foresters may have a more extensive list of forestry consultants in your area, since not all consultants belong to the national association.

The **Forest Resource Center** is a nonprofit corporation that conducts educational programs for private woodland owners, youth, and others. Its programs are aimed at improving management and utilization of hardwood forest resources in southeastern Minnesota and adjoining areas of Wisconsin and Iowa. The Center is developing a demonstration forest where a variety of forest regeneration and management practices can be seen, including how to grow shiitake mushrooms. It offers workshops on forestry subjects in cooperation with public agencies. Contact Forest Resource Center, 1991 Brightsdale Road, Route 2, Box 156A, Lanesboro, MN 55949.

ASSOCIATIONS

Each state has forestry-related associations for woodland owners. For information about them contact your state forester or extension service. Below are descriptions of national forestry associations.

American Forests is America's oldest national conservation organization focused on trees and forests. It is independent and nonpartisan. It advances the intelligent management, enjoyment, and use of forests, soil, water, wildlife, and other natural resources, both rural and urban. It promotes an enlightened public appreciation of natural resources and their role in the social, recreational, and economic life of the nation's woodlands and communities. It does so through publication of its magazine, *American Forests*; its newsletters, the *Urban Forest Forum* and *Resource Hotline*; and through its Global ReLeaf program, which promotes the growing of more and better trees for America and the world. Contact the American Forestry Association, P.O. Box 2000, Washington, DC 20013.

The American Tree Farm System is a national organization of woodland owners. Each member receives a forest management plan developed for his or her particular woodland by a professional forester. A forester will inspect the woodland every five years and update the forest management plan accordingly. Members are eligible to compete for prizes in "Outstanding Tree Farmer" contests and may subscribe to *The American Tree Farmer* magazine. In return the member agrees to follow the plan and protect the woodland from fire, insects, disease, and grazing. This service is offered at no cost. Contact your state forester or the American Forest Council, 1250 Connecticut Avenue NW, Washington, DC 20036.

The National Woodland Owners Association (NWOA) is an organization of nonindustrial private woodland owners with members from all 50 states. It is affiliated with state and county woodland owner associations throughout the United States. NWOA is independent of the forest products industry and forestry agencies, and works with all organizations to promote nonindustrial forestry and the best interests of woodland owners. Members receive four issues of *National Woodlands* magazine and eight issues of *Woodland Report* with late-breaking forestry news from Washington, D.C., and state governments. An introductory visit from a professional forester is available in most states (for holdings of 20 acres or larger), plus other member benefits. Contact National Woodland Owners Association, 374 Maple Avenue East, Suite 210, Vienna, VA 22180.

The Nature Conservancy is an international nonprofit conservation organization. It acquires land with unique plants, animals, or habitats and usually transfers it to federal, state, or local agencies that can manage land for public use. The conservancy also conducts research on unique plants, animals, and habitats. Members can participate in a wide range of educational outings. Contact The Nature Conservancy, 1313 5th St. S.E., Box 110, Minneapolis, MN 55414.

The **Society of American Foresters (SAF)** is a professional society composed of graduates of professional forestry curricula in SAF-accredited schools, forestry students, scientists or practitioners in fields closely allied to forestry who hold bachelor or higher degrees, and students or graduates of SAF-recognized forestry technician programs. The society's *Journal of Forestry* keeps members up to date on forestry issues, practical forestry information, and information on new forestry products and equipment. Members may attend SAF chapter, state, and national educational meetings. SAF has produced a publication and video that describe forestry careers. Contact the Society of American Foresters, 5400 Grosvenor Lane, Bethesda, MD 20814.

The **Ruffed Grouse Society** is a national nonprofit conservation organization dedicated to enhancing habitat for ruffed grouse, American woodcock, and other wildlife that utilize young forests. Society-supported research has yielded publications useful for both public land managers and private woodland owners. Members receive *RGS* magazine and notices about chapter meetings and fundraising banquets. Habitat management publications are available. Contact Ruffed Grouse Society, 1400 Lee Drive, Coraopolis, PA 15108.

The **Walnut Council** sponsors meetings of agencies, landowners, producers, and woodland users interested in walnut culture and utilization. The council also functions as a means to exchange information and propagational materials. Members are kept up to date on the best known techniques to grow walnut by means of the *Walnut Council Bulletin*, a quarterly newsletter. Contact the Walnut Council, 5603 West Raymond Street, Indianapolis, IN 46241.

APPENDIX B FORESTRY MEASUREMENTS AND CONVERSIONS

LAND MEASUREMENTS

Linear

1 link = 7.92 inches

100 links = 1 chain

1 chain = 66 feet

80 chains = 1 mile

1 rod = 16.5 feet

4 rods = 1 chain

1 mile = 5,280 feet

Square

1 square rod = 272.25 square feet

10 square chains = 1 acre

1 acre = 43,560 square feet; a square acre measures about 209 feet on each side

1 square mile (section) = 640 acres

1 township = 36 sections

TREE AND LOG MEASUREMENTS

1 board foot = 144 cubic inches of solid wood

1 cubic foot = 1,728 cubic inches of solid wood or approximately 5 to 6 board feet because of losses in sawing

1 cunit = 100 cubic feet of solid wood

1 cord = a closely stacked pile of wood containing 128 cubic feet of wood, bark, and air spaces. Usually measured as 4 feet x 4 feet x 8 feet. Usually contains 70 to 90 cubic feet of solid wood or approximately 500 board feet.

APPROXIMATE ENGLISH - METRIC CONVERSIONS

English to Metric

1 inch = 2.540 centimeters

1 foot or 12 inches = 30.480 centimeters

1 yard or 3 feet = 0.914 meters

1 U.S. statute mile or 5,280 feet = 1.609 kilometers

1 acre or 43,560 square feet = 0.405 hectares

1 cubic foot or 1,728 cubic inches = 0.028 cubic meters

Metric to English

1 centimeter or 10 millimeters = 0.394 inches

1 meter = 39.370 inches

1 kilometer or 1,000 meters = 0.621 U.S. statute mile

1 hectare or 10,000 square meters = 2.471 acres

1 cubic meter or 1,000,000 cubic centimeters = 35.315 cubic feet

APPENDIX C
SITE INDEX CURVES FOR SELECTED TREE SPECIES

INTERPRETING SITE INDEX CURVES

Site index is the height to which trees will grow over a given period—usually 50 years in the Midwest. Trees are expected to grow taller on good sites than on poor sites. Site index curves can be used to estimate relative site quality from the average age and average total height of dominant and codominant trees on a particular site. Each tree species has its own set of site index curves.

To use the site index curves in this appendix, first measure the age and total height of several dominant and codominant trees of a single species on a site. Calculate the average total age and average total height of those sample trees. Then refer to the site index curves for that species.

For example, refer to the site index curves for quaking aspen in Appendix C-1. If the average age of aspen trees in your stand of interest is 40 years (horizontal axis) and the average total height is 62 feet (vertical axis), then those two lines converge closest to the site index curve for 70. This means that aspen trees in your stand are expected to be 70 feet tall when they are 50 years old. They are growing on a good site.

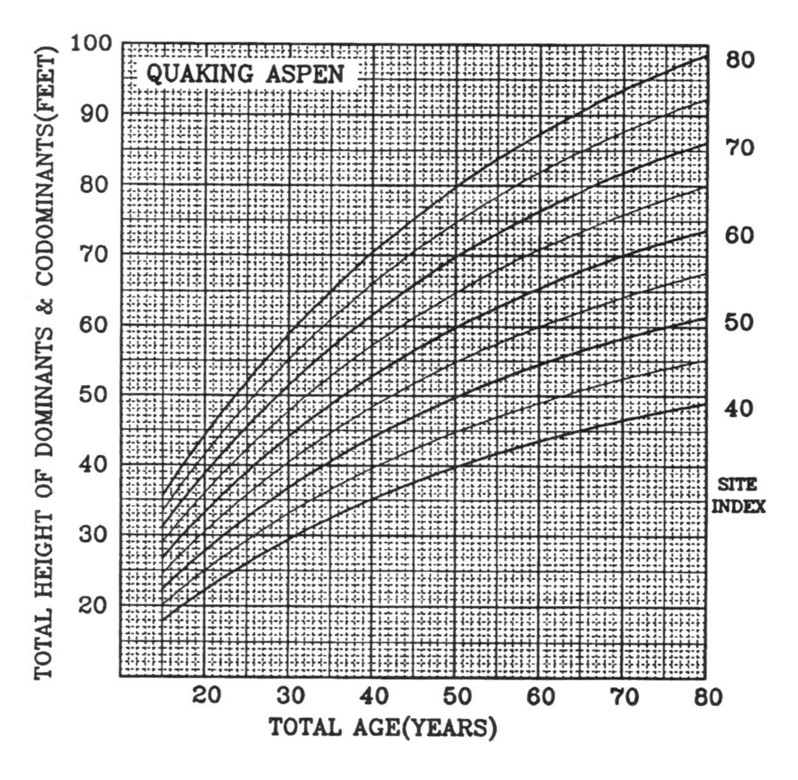

C-1: Quaking Aspen

Note: Add 4 years to age at DBH to obtain total age.

SOURCE: Carmean W. H., J. T. Hahn, and R. D. Jacobs. 1989. *Site Index Curves for Forest Tree Species in the Eastern United States (General Technical Report NC-128).* USDA Forest Service, North Central Forest Experiment Station, 1992 Folwell Avenue, St. Paul, MN 55108. p. 46.

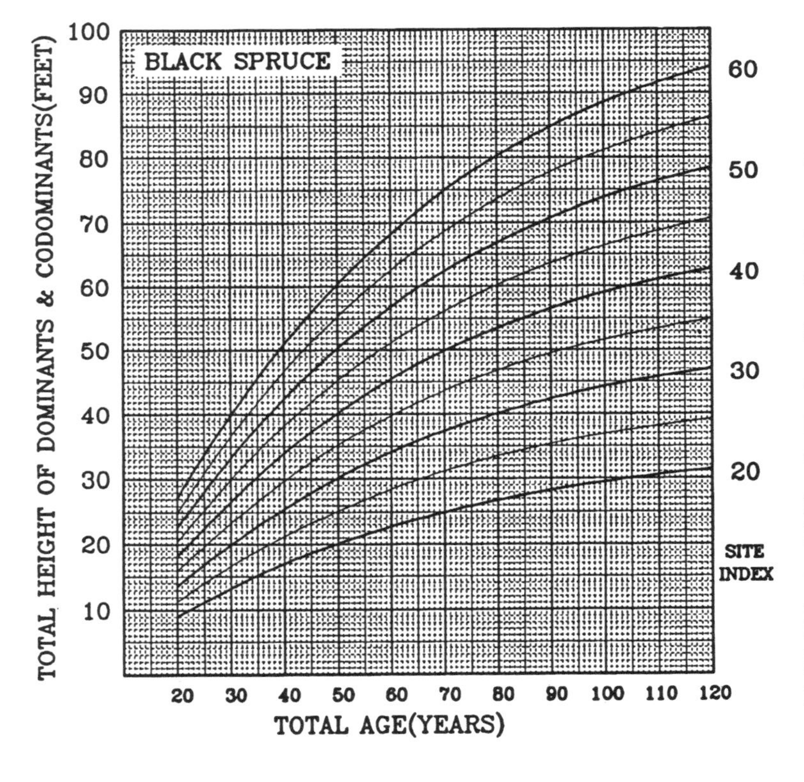

C-2: Black Spruce

Note: To obtain total age, add years to age at DBH according to the table below.

Site Index:	20	30	40	50	60	70	80	90
Add Years:	15	13	11	10	9	8	7	6

Note: Curves are based on plots in northeastern Minnesota—Superior National Forest.

SOURCE: Carmean W. H., J. T. Hahn, and R. D. Jacobs. 1989. *Site Index Curves for Forest Tree Species in the Eastern United States (General Technical Report NC-128).* USDA Forest Service, North Central Forest Experiment Station, 1992 Folwell Avenue, St. Paul, MN 55108. p. 85.

C-3 Black Walnut Plantations

Note: Determine total age from stumps or planting records; do not damage trees by using an increment borer.

Note: Curves are based on plots in southern Illinois.

SOURCE: Carmean W. H., J. T. Hahn, and R. D. Jacobs. 1989. *Site Index Curves for Forest Tree Species in the Eastern United States (General Technical Report NC-128).* USDA Forest Service, North Central Forest Experiment Station, 1992 Folwell Avenue, St. Paul, MN 55108. p. 33.

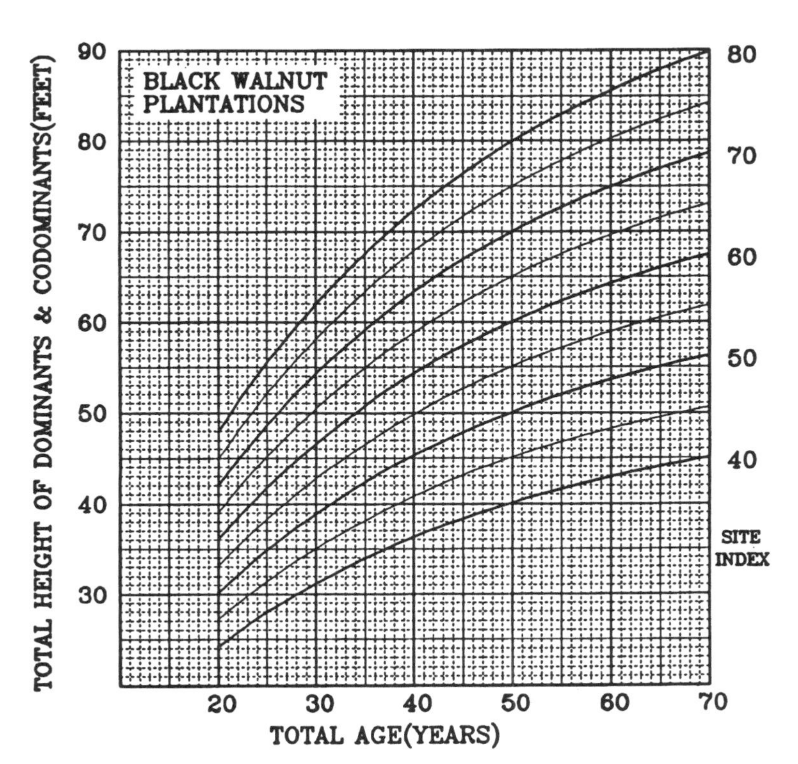

C-4: Green Ash

Note: Add 2 years to age at DBH to obtain total age.

Note: Curves are based on plots in Mississippi Valley alluvium in Louisiana, Mississippi, Arkansas, and Tennessee.

SOURCE: Carmean W. H., J. T. Hahn, and R. D. Jacobs. 1989. *Site Index Curves for Forest Tree Species in the Eastern United States (General Technical Report NC-128).* USDA Forest Service, North Central Forest Experiment Station, 1992 Folwell Avenue, St. Paul, MN 55108. p. 30.

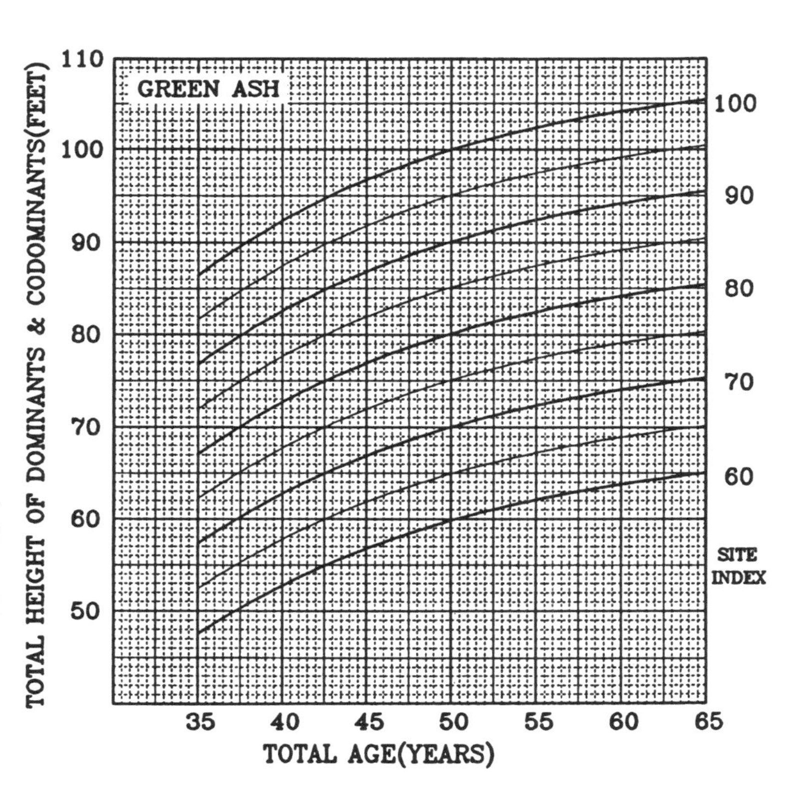

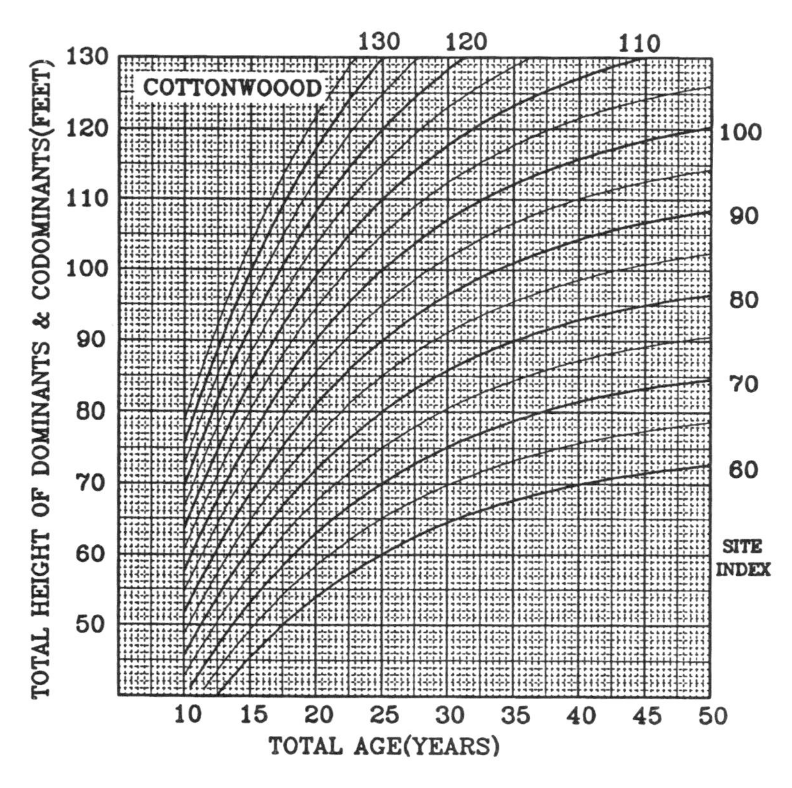

C-5: Eastern Cottonwood

Note: Add 2 years to age at DBH to obtain total age.

Note: Curves are based on plots in Iowa bottomlands.

SOURCE: Carmean W. H., J. T. Hahn, and R. D. Jacobs. 1989. *Site Index Curves for Forest Tree Species in the Eastern United States (General Technical Report NC-128).* USDA Forest Service, North Central Forest Experiment Station, 1992 Folwell Avenue, St. Paul, MN 55108. p. 45.

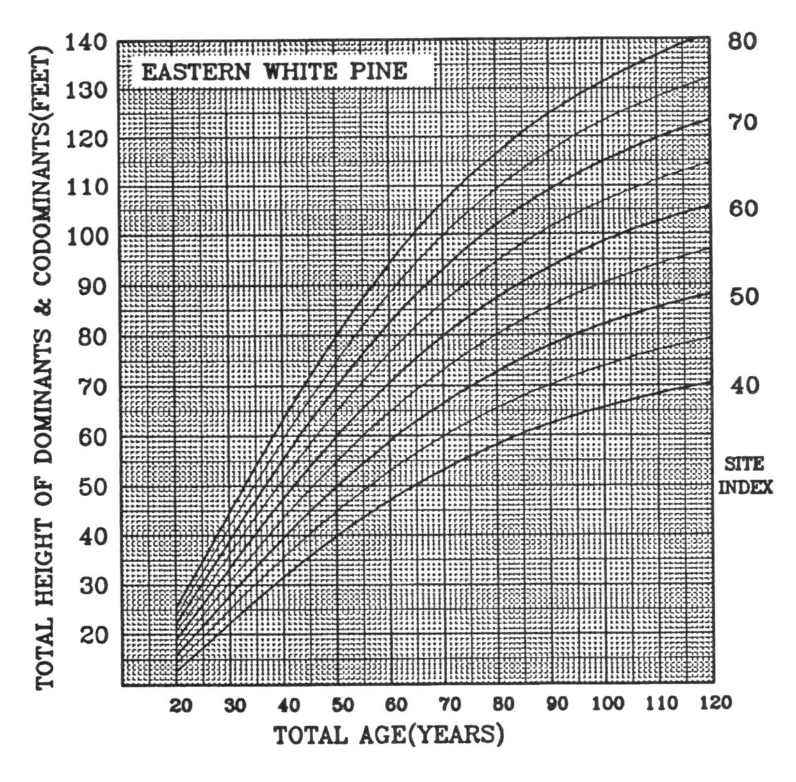

C-6: Eastern White Pine

Note: To obtain total age, add years to age at DBH according to the table below.

Site Index:	40	50	60	70	80
Add Years:	12	12	10	8	6

Note: Curves are based on plots in northern Wisconsin.

SOURCE: Carmean W. H., J. T. Hahn, and R. D. Jacobs. 1989. *Site Index Curves for Forest Tree Species in the Eastern United States (General Technical Report NC-128).* USDA Forest Service, North Central Forest Experiment Station, 1992 Folwell Avenue, St. Paul, MN 55108. p. 118.

C-7: Jack Pine

Note: To obtain total age, add years to age at DBH according to the table below.

Site Index:	30	40	50	60	70	80	90
Add Years:	9	8	7	6	5	4	4

Note: Curves are based on plots in the Lake States.

SOURCE: Carmean W. H., J. T. Hahn, and R. D. Jacobs. 1989. *Site Index Curves for Forest Tree Species in the Eastern United States (General Technical Report NC-128).* USDA Forest Service, North Central Forest Experiment Station, 1992 Folwell Avenue, St. Paul, MN 55108. p. 89.

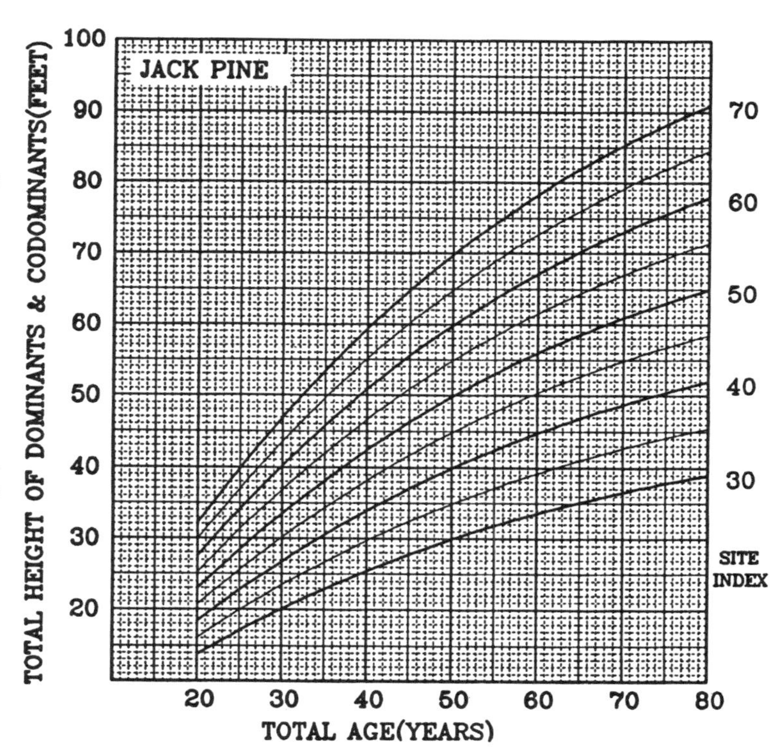

C-8: Hardwood Comparisons

Note: If you know the site index for one species, you can determine the site index for another species on the same site by moving vertically up or down to the curve for each species of interest. For example, if you know the site index for aspen is 72, find that number on the aspen curve. Then read directly downward to the white ash curve to find the corresponding site index of 68 for white ash.

SOURCE: USDA, Forest Service. 1985. *Northern Hardwood Notes (Note 4.02).* U.S. Government Printing Office, Washington, DC 20402.

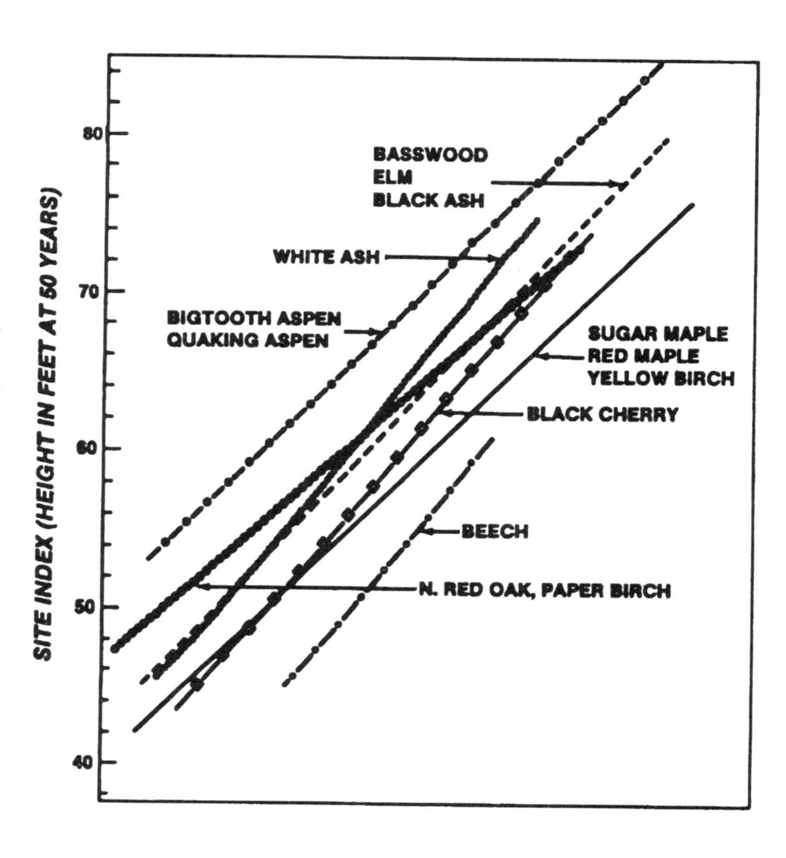

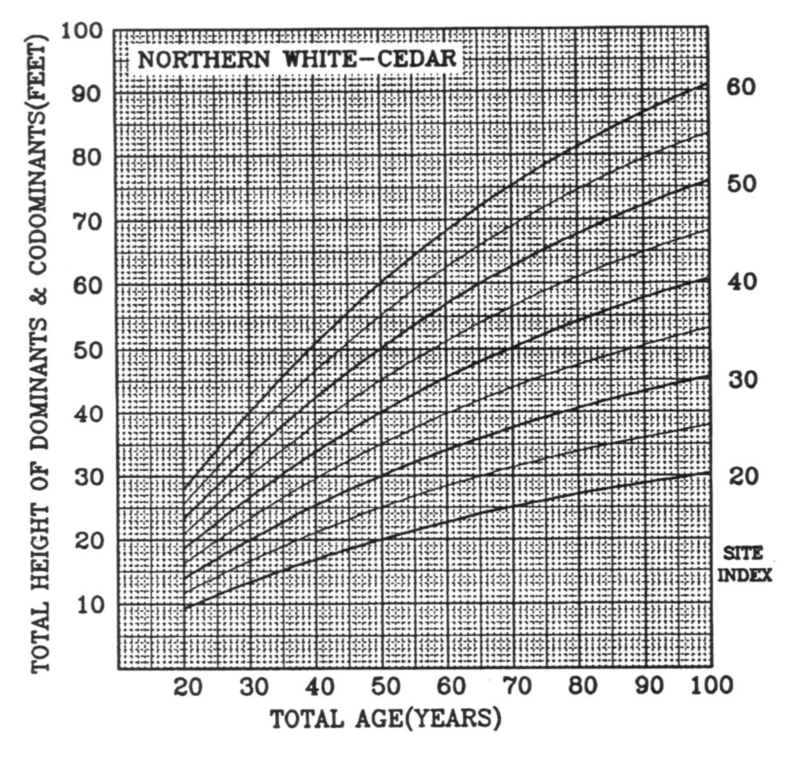

C-9: Northern White-Cedar

Note: To obtain total age, add years to age at DBH according to the table below.

Site Index: 20 30 40 50 60

Add Years: 20 15 15 10 10

Note: Curves are based on plots in the Lake States.

SOURCE: Carmean W. H., J. T. Hahn, and R. D. Jacobs. 1989. *Site Index Curves for Forest Tree Species in the Eastern United States (General Technical Report NC-128).* USDA Forest Service, North Central Forest Experiment Station, 1992 Folwell Avenue, St. Paul, MN 55108. p. 141.

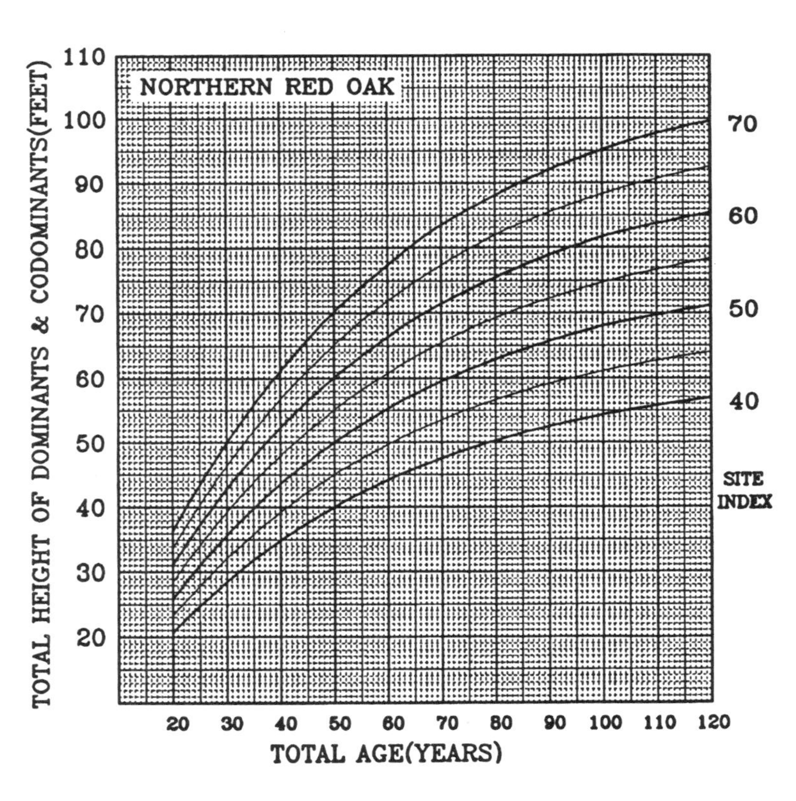

C-10: Northern Red Oak

Note: Add 4 years to age at DBH to obtain total age.

Note: Curves are based on plots in southwestern Wisconsin.

SOURCE: Carmean W. H., J. T. Hahn, and R. D. Jacobs. 1989. *Site Index Curves for Forest Tree Species in the Eastern United States (General Technical Report NC-128).* USDA Forest Service, North Central Forest Experiment Station, 1992 Folwell Avenue, St. Paul, MN 55108. p. 62

C-11: Red (Norway) Pine

Note: To obtain total age, add years to age at DBH according to the table below.

Site Index:	40	50	60	70+
Add Years:	8	6	5	4

Note: Curves are based on plots in Minnesota.

SOURCE: Carmean W. H., J. T. Hahn, and R. D. Jacobs. 1989. *Site Index Curves for Forest Tree Species in the Eastern United States (General Technical Report NC-128).* USDA Forest Service, North Central Forest Experiment Station, 1992 Folwell Avenue, St. Paul, MN 55108. p. 110.

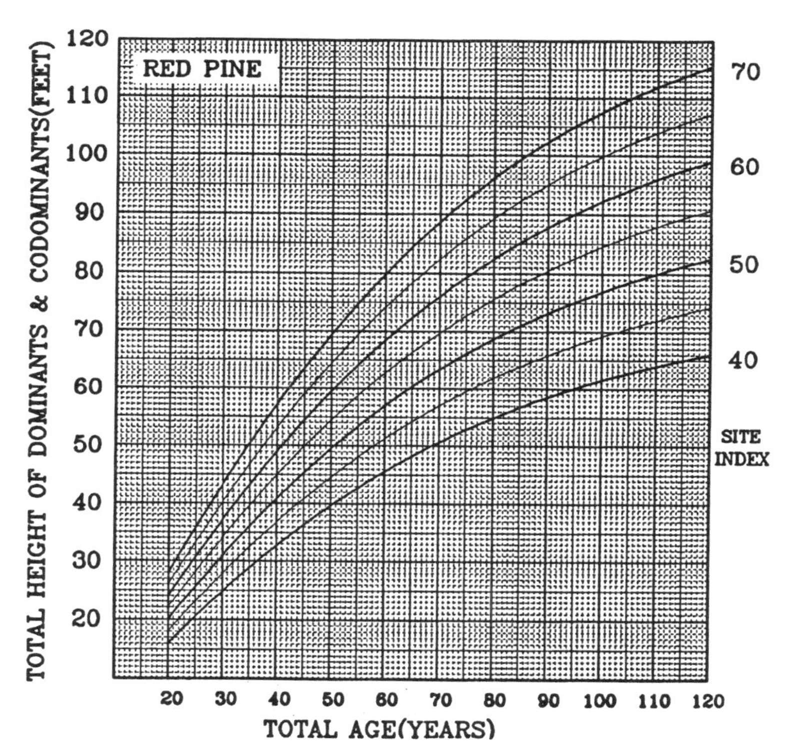

C-12: White Spruce

Note: These curves assume that age at DBH is 4.5 years.

Note: Curves are based on plots in Minnesota.

Source: Carmean W. H., J. T. Hahn, and R. D. Jacobs. 1989. *Site Index Curves for Forest Tree Species in the Eastern United States (General Technical Report NC-128).* USDA Forest Service, North Central Forest Experiment Station, 1992 Folwell Avenue, St. Paul, MN 55108. p. 81.

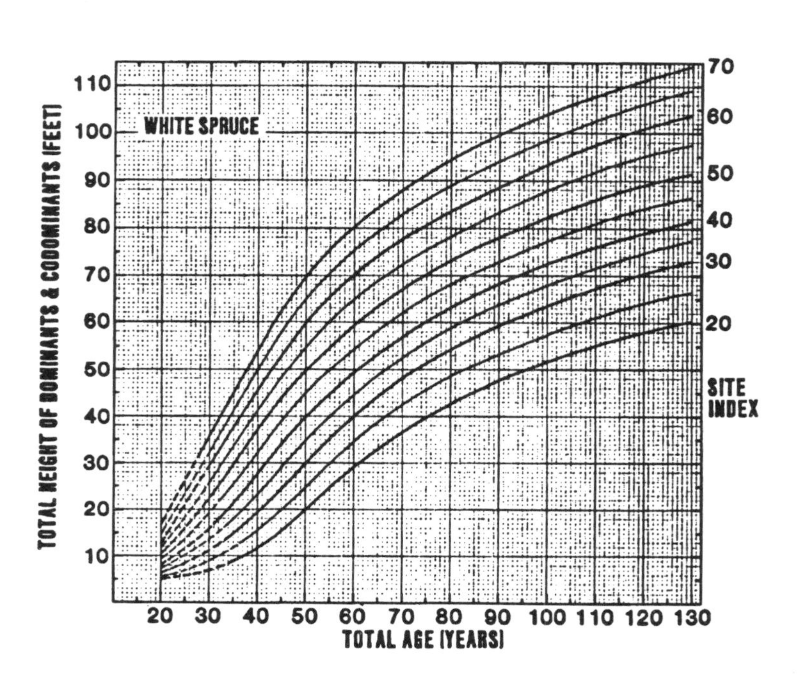

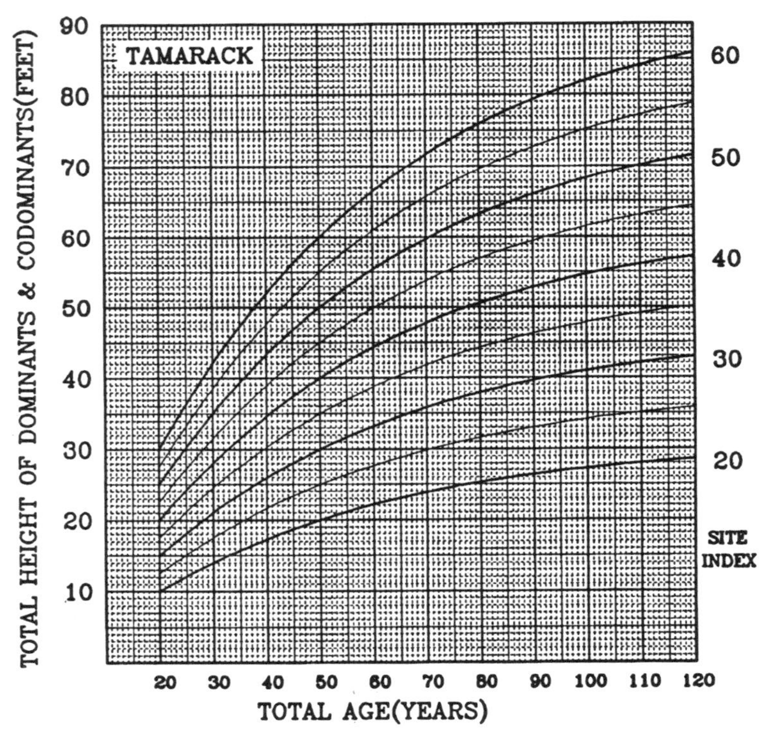

C-13:Tamarack

Note: To obtain total age, add years to age at DBH according to the table below.

Site Index:	20	30	40	50-90
Add Years:	12	10	7	5

Note: Curves are based on plots in Minnesota.

SOURCE: Carmean W. H., J. T. Hahn, and R. D. Jacobs. 1989. *Site Index Curves for Forest Tree Species in the Eastern United States (General Technical Report NC-128).* USDA Forest Service, North Central Forest Experiment Station, 1992 Folwell Avenue, St. Paul, MN 55108. p. 75.

APPENDIX D STOCKING CHARTS FOR SELECTED TREE SPECIES AND FOREST TYPES

D-1: Elm-Ash-Cottonwood

D-2: Nearly Pure Even-Aged Eastern White Pine

D-3: Jack Pine

D-4: Even-Aged Management of Northern Hardwoods

D-5: Upland Central Hardwoods

D-6: Red (Norway) Pine

D-7: Even-Aged Spruce-Balsam Fir Stands

INTERPRETING STOCKING CHARTS

Stocking charts are useful thinning guides. If you know the square feet of basal area and number of trees per acre in a stand, you can refer to a stocking chart for the species of interest and determine whether the stand is overstocked, fully stocked, or understocked. Stands above the **A** level on a stocking chart are overstocked and should be thinned back to near the **B** level to increase tree growth rate.

For example, refer to Appendix D-1. If your stand had a basal area of 110 square feet and 200 trees per acre, it would be at the **A** level where it is nearly overstocked. Trees in the stand would grow faster if the stand were thinned. Trees in this sample stand have an average stand diameter of 10 inches. Follow the line for 10 inches diameter down to the **B**-level curve. It intersects the **B**-level curve where the basal area is 68 square feet and there are 125 trees per acre. The residual trees would grow best if the stand were thinned back to this stocking level.

However, a stand that is heavily thinned may be subject to windthrow and epicormic branching. As a rule of thumb, do not remove more than one-third of the basal area from a stand at any one time. Applying this principal to the example above, the stand should be thinned down to 74 square feet of basal area and approximately 135 trees per acre.

D-1: Elm-Ash-Cottonwood

SOURCE: Myers, C. C. and R. G. Buchman. 1984. *Manager's Handbook for Elm-Ash-Cottonwood in the North Central States* (General Technical Report NC-98). USDA Forest Service, North Central Forest Experiment Station, 1992 Folwell Avenue, St. Paul, MN 55108. p. 9.

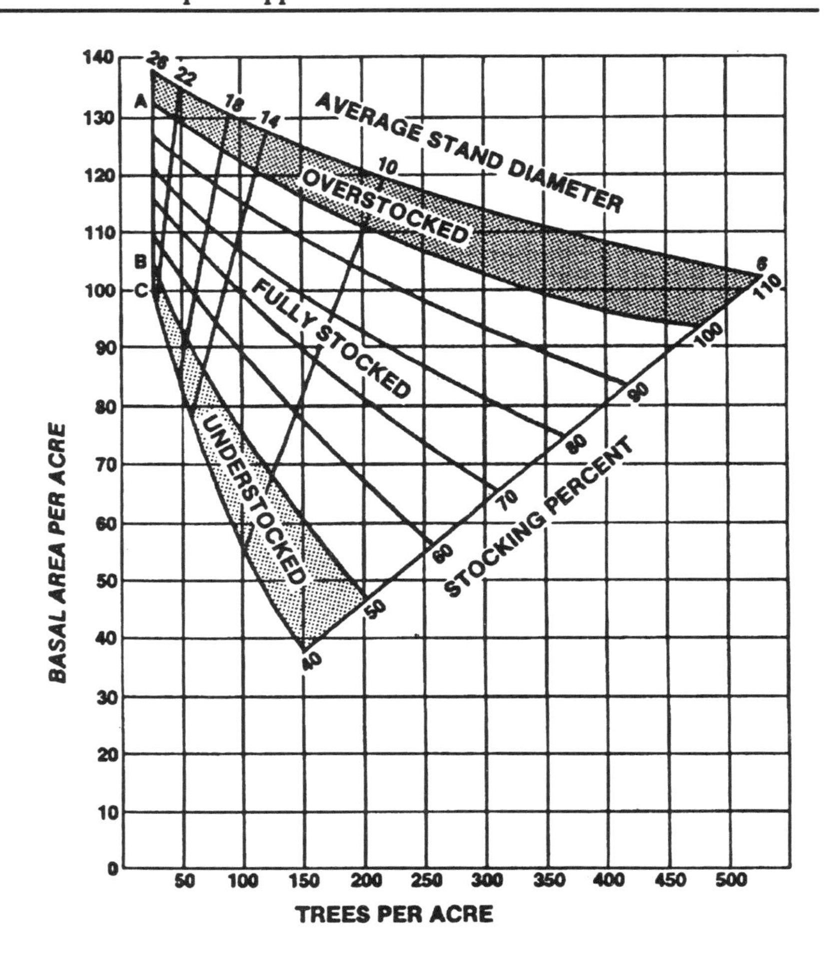

D-2: Nearly Pure Even-Aged Eastern White Pine

SOURCE: U.S. Department of Agriculture, Forest Service. 1990. *Silvics of North America, Volume I Conifers (Agricultural Handbook No. 654).* U.S. Government Printing Office, Washington, DC 20402. p. 482.

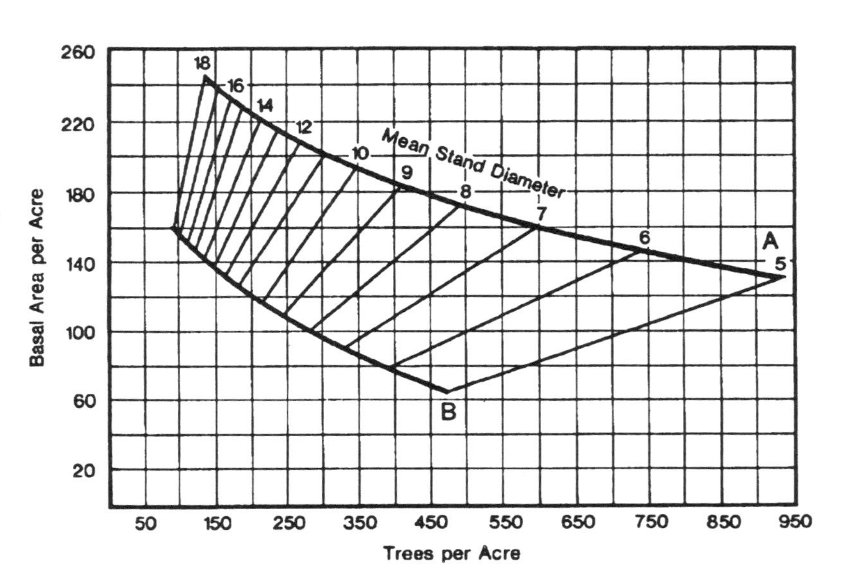

D-3: Jack Pine

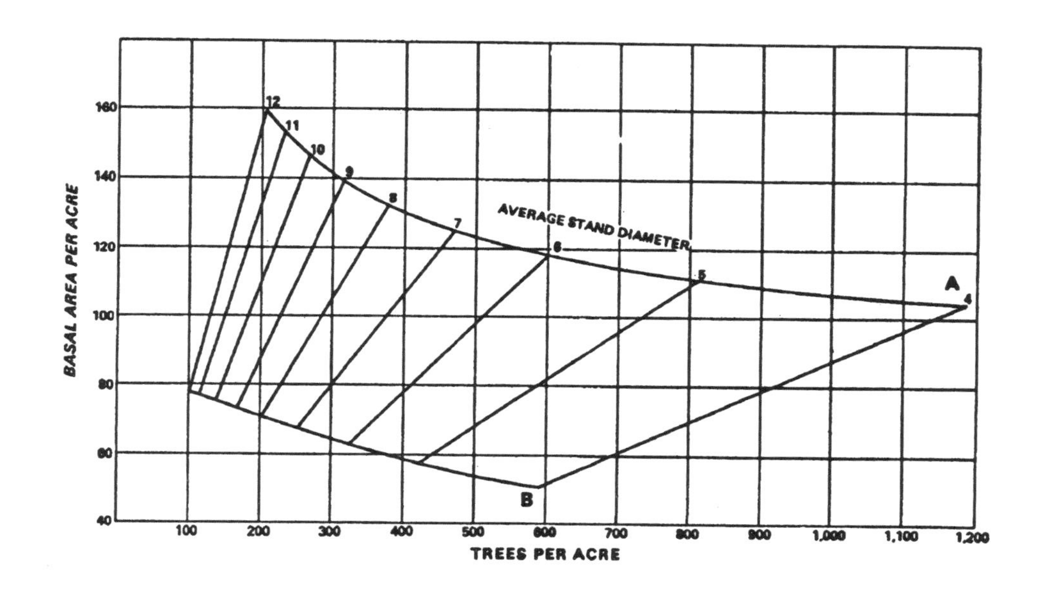

SOURCE: Benzie, J. W. 1977. *Manager's Handbook for Jack Pine in the North Central States (General Technical Report NC-32).* USDA Forest Service, North Central Forest Experiment Station, 1992 Folwell Avenue, St. Paul, MN 55108. p. 11.

D-4: Even-Aged Management of Northern Hardwoods

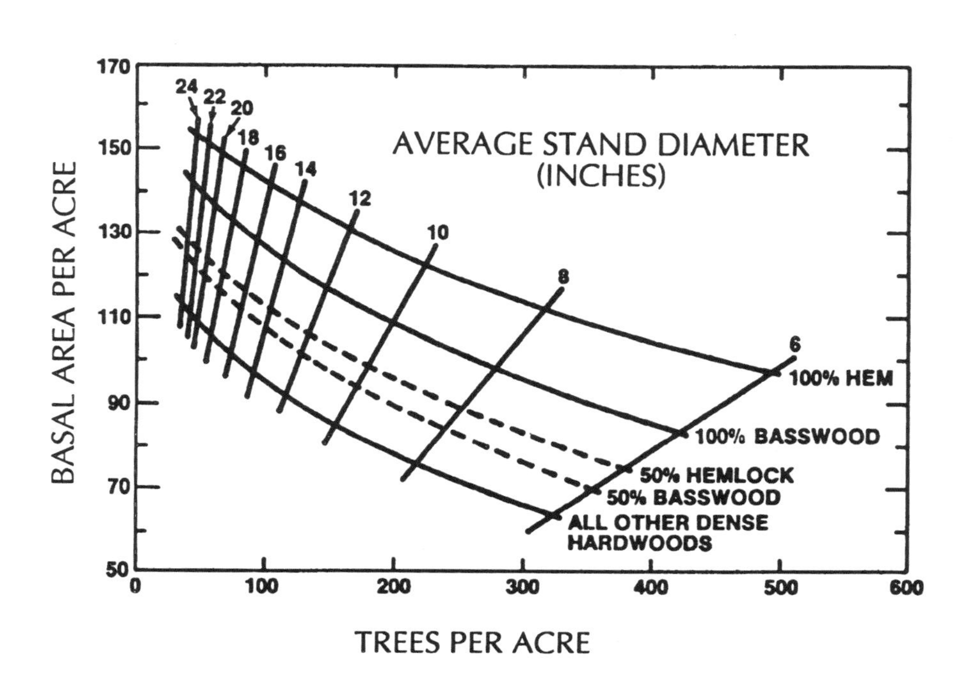

Note: Find the current basal area per acre of your stand on the vertical axis and the number of trees per acre on the horizontal axis. Where these lines meet is the average tree diameter. Next extend a line paralleling the average tree diameter line to the appropriate species curve. (Use the hemlock and basswood curves labeled 100 percent only if the stand you are thinning is at least 80 percent stocked with that particular species. Use the dashed curves for stands that are about 50 percent stocked with that particular species.) Go horizontally from the intersection of the average tree diameter line and species curve to the vertical axis and read the residual basal area that is optimum for that stand.

SOURCE: U.S. Department of Agriculture, Forest Service. 1985. *Northern Hardwood Notes (Note 4.03).* U.S. Government Printing Office, Washington, DC 20402.

D-5: Upland Central Hardwoods

Note: For average tree diameters of 7 to 15 inches, use Chart A.

For average tree diameters of 3 to 7 inches, use Chart B.

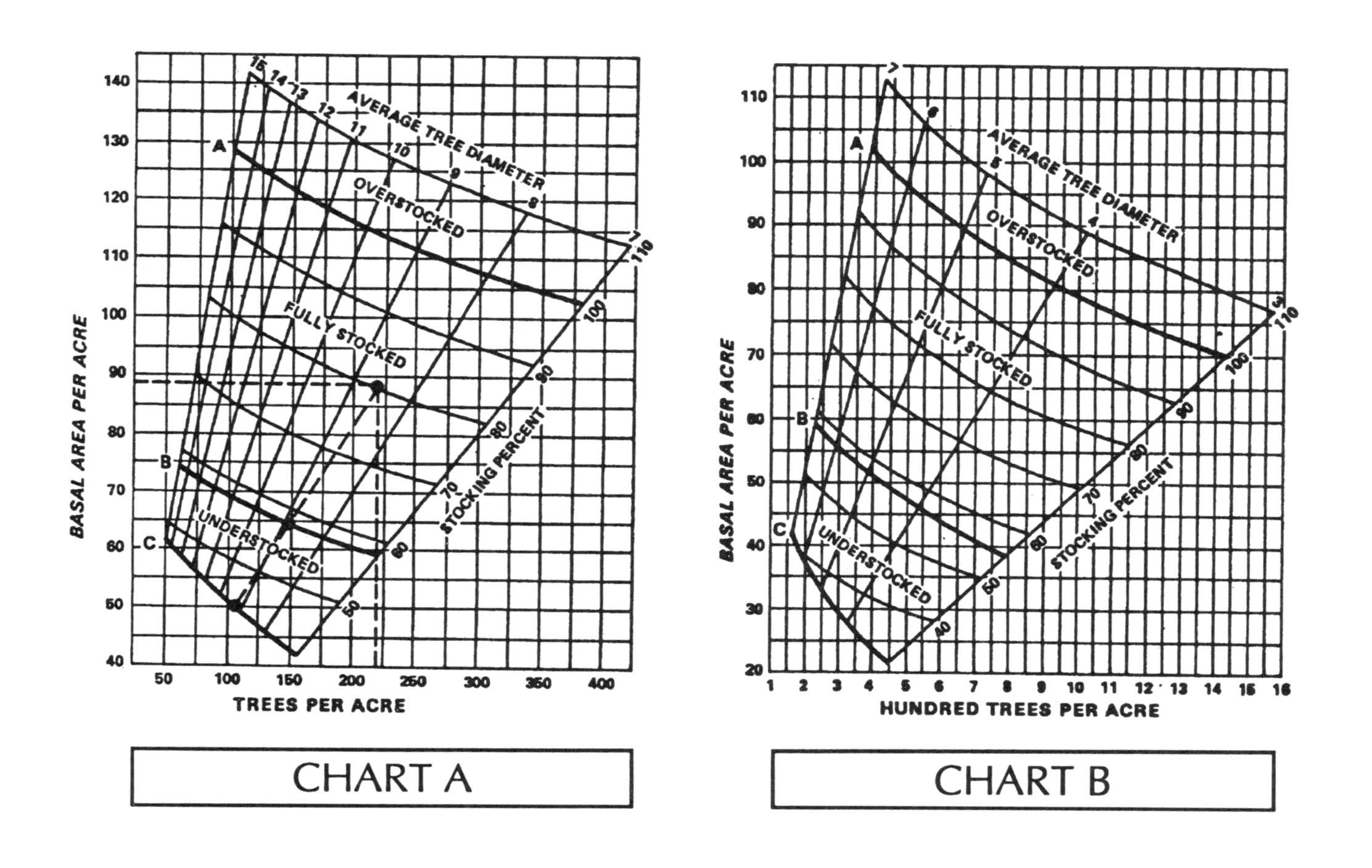

SOURCE: Sander, I. L. 1977. *Manager's Handbook for Oaks in the North Central States (General Technical Report NC-37).* USDA Forest Service, North Central Forest Experiment Station, 1992 Folwell Avenue, St. Paul, MN 55108. p. 29.

D-6: Red (Norway) Pine

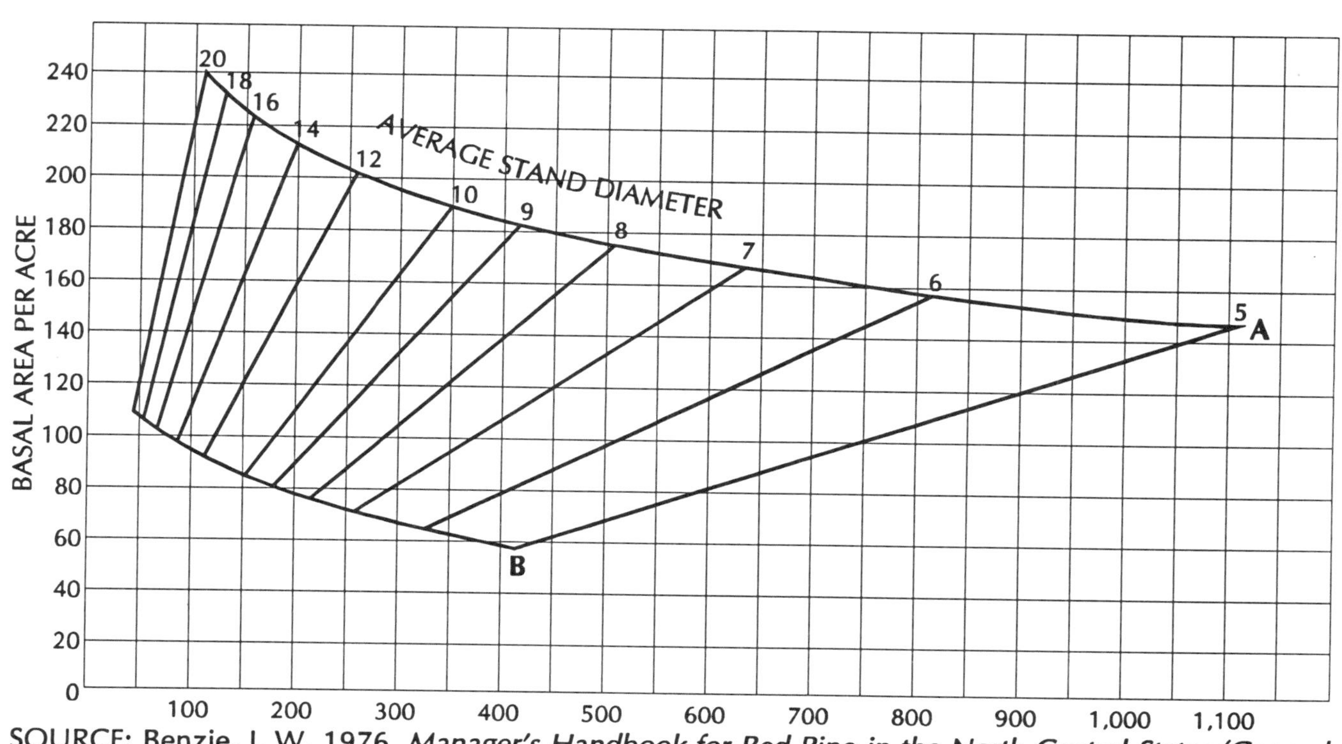

SOURCE: Benzie, J. W. 1976. *Manager's Handbook for Red Pine in the North Central States (General Technical Report NC-33).* USDA Forest Service, North Central Forest Experiment Station, 1992 Folwell Avenue, St. Paul, MN 55108. p. 13.

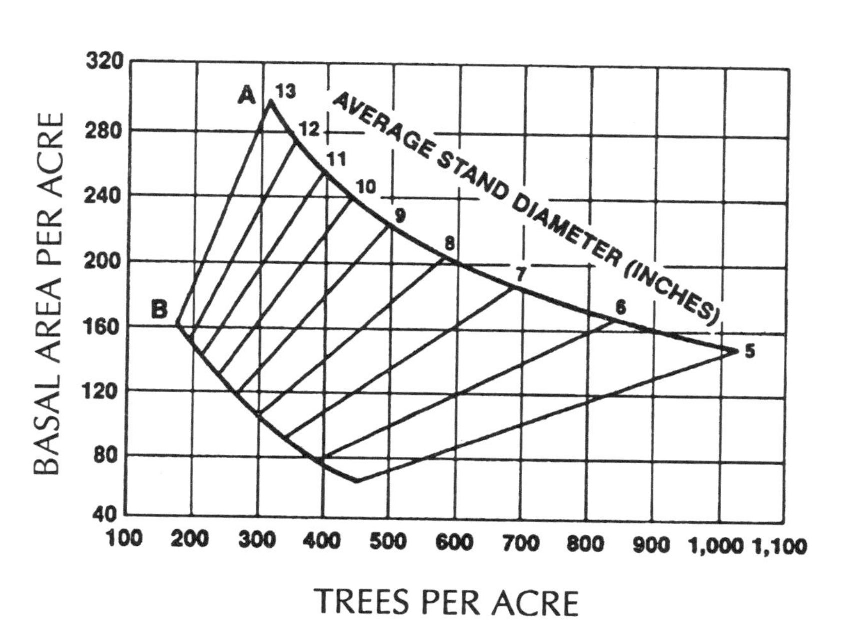

D-7: Even-Aged Spruce-Balsam Fir Stands

SOURCE: Johnston, W. F. 1986. *Manager's Handbook for Balsam Fir in the North Central States (General Technical Report NC-111).* USDA Forest Service, North Central Forest Experiment Station, 1992 Folwell Avenue, St. Paul, MN 55108. p. 8.

GENERAL REFERENCES

Beattie, M., C. Thompson, and L. Levine. 1983. *Working With Your Woodland: A Landowner's Guide.* University Press of New England, Hanover, NH 03755. 310 pp.

Canada Department of Regional Economic Expansion and Nova Scotia Department of Lands and Forests. 1980. *The Trees Around Us: A Manual for Good Forest Practice for Nova Scotia.* 206 pp.

Forbes, R. D. 1976. *Woodlands for Profit and Pleasure (2nd ed.).* American Forestry Association, P.O. Box 2000, Washington, DC 20077. 169 pp.

Minckler, L. S. 1980. *Woodland Ecology: Environmental Forestry for the Small Owner (2nd ed.).* Syracuse University Press, Syracuse, NY 13210. 241 pp.

Minnesota Association of Soil and Water Conservation Districts. *Minnesota Tree Handbook.* 2095 Delaware Avenue, Mendota Heights, MN 55108. 408 pp.

Stoddard, C. H. and G. M. Stoddard. 1987. *Essentials of Forestry Practice (4th Ed.).* John Wiley & Sons, New York. 451 pp.

USDA Forest Service. 1974. *Seeds of Woody Plants in the United States (Agriculture Handbook No. 450).* U.S. Government Printing Office, Washington, DC 20402. 883 pp.

GLOSSARY

Abandonment Test — An investment monitoring procedure that calculates an abandonment value and updated NPV, then compares these two values.

Acre — An area of land that contains 43,560 square feet.

Activity — An action, cost, or benefit related to an investment alternative (e.g., land purchase, planting, annual property taxes).

After-Tax — Values in a cash flow table that include income tax effects.

Age Class — A period of years (commonly one to five) during which a stand regenerates.

All-Aged Forest — A forest or stand that contains trees of all, or almost all, age classes. Also called "uneven-aged" forest. This is in contrast to an "even-aged" forest.

Amortization — The process by which the basis of certain assets (e.g.,reforestation expenses for timber production) is deducted from income taxes over a fixed period of years.

Annual Percentage Rate (APR) — The interest rate that is actually paid or earned. It considers multiple compounding periods within a year.

Artificial Reproduction (Artificial Regeneration) — *See Reproduction.*

Aspect — the compass direction towards which a slope faces.

Basal Area

- of a tree: the cross-sectional area (in square feet) of the trunk at breast height (4-1/2 feet above the ground). For example, the basal area of a tree 14 inches DBH is approximately 1 square foot.

- of an acre of forest: the sum of basal areas of the individual trees on the acre. A well-stocked northern hardwood stand might contain 80 to 100 square feet of basal area per acre.

Basis — In general, the acquisition cost for property. However, the basis for property acquired by gift or inheritance is not its cost (see Chapter 12). The **original basis** is the basis at the time the property was first acquired. The **adjusted basis** is the original basis less any reductions made because of depreciation, depletion, amortization, or losses claimed on income taxes, plus any additions made by capitalization of improvement expenses, carrying charges, or additions to the asset.

Before-Tax — Values in a cash flow table that do not include income tax effects.

Benefit — A good and/or service produced as a result of having implemented an investment alternative. A benefit must have some human use or value to be included in a financial analysis.

Best Alternative(s) — The alternative or alternatives that most closely meet the overall objectives of the investor, considering both financial and nonfinancial criteria.

Block — An area of woodland based on geographic boundaries, tree species, timber products, accessibility, or other timber characteristics.

Board Foot — A unit for measuring wood volumes equaling 144 cubic inches. It commonly is used to measure and express the amount of wood in a tree, sawlog, veneer log, or individual piece of lumber. For example, a piece of wood 12 inches x 12 inches x 1 inch and one measuring 12 inches x 3 inches x 4 inches each contains 1 board foot of wood.

Bole — The main trunk of a tree.

Bolt — A short log or a squared timber cut from a log, commonly 8 feet long.

Branch Collar — A natural swelling at the base of a tree limb (see Figure 36).

Broadcast Burn — A controlled fire that is set purposely to burn across an area and eliminate woody debris and undesirable small trees and shrubs.

Browse — Growing leaves, shoots, or twigs used as animal fodder. As a verb, to feed on leaves, shoots, or twigs.

Buck — The process of cutting a tree stem into logs.

Burl — A growth of woody tissue protruding outward from a tree stem. Usually rounded in shape.

Canopy — The layer of vegetation in a forest made up of tree crowns.

Capital — Money and other property used in transacting a business or daily operations.

Capital Account — An account used to record the basis and quantity of capital assets such as land, timber, buildings, and equipment.

Capital Asset — In general, everything you own and use for personal purposes or investment. It does not include property held mainly for sale to customers or accounts receivable in the ordinary course of a trade or business or for services rendered as an employee. Depreciable property and real property used in your trade or business are not capital assets, but may be treated as such under certain circumstances.

Capital Expense — An expense incurred to acquire a capital asset or to make significant repairs or improvements to it.

Cash Flow — The individual monetary costs and benefits for an investment alternative.

Cash Flow Table — Graphic display of activity cash flows by year throughout the life of an investment alternative. Total annual costs and benefits, as well as the net benefit or loss for each year, also are shown.

Chain — A distance of 66 feet.

Cleaning — *See Release Cutting.*

Clearcutting Harvest — A harvesting method which removes all the trees on an area in one operation. Regeneration occurs from seed or seedlings present before cutting, from dormant seed on the ground, from seed that disperses from adjoining stands, or from artificial planting or seeding. Clearcutting is used most often with species that require full sunlight to reproduce and grow well. Produces an even-aged forest stand.

Clone — A plant group derived from a single individual through asexual (vegetative) reproduction.

Codominant — *See Crown Classification.*

Commercial Cut — A timber harvest that yields a net income (receipts for the sale of products exceed the cost of the cutting).

Compounding — The process of determining the future value of a present payment.

Conifer — A tree belonging to the order *Coniferales* which usually is evergreen, cone-bearing, and has leaves that are needle-, awl- or scale-like, such as pine, spruce, fir and cedar; often referred to as a "softwood."

Conservation — The protection, improvement, and wise use of natural resources to assure the attainment of their highest economic and/or social values.

Cord — A stack of logs containing 128 cubic feet. Normal dimensions of a standard cord are 4 feet x 4 feet x 8 feet. In the Lake States, pulpwood cords usually are 4 feet x 4 feet x 100 inches.

Cost — An activity that is put into an investment to achieve a result or benefit. Each investment generally requires several cost activities.

Crop Tree — A tree that is to be grown to maturity and that is not removed from the forest before the final harvest. Usually selected on the basis of its species, location with respect to other trees, and quality.

Crown — The leaves and branches of a tree.

Crown Classification — Ranking of individual trees in a stand according to the relative size and height of their crowns. In descending order of crown height and size, the classes commonly used are **dominant**, **codominant**, **intermediate**, and **suppressed**.

Crown Ratio — Percentage of total tree height that is occupied by living branches.

Cruise — Process of collecting timber inventory information such as tree volumes.

Cubic Foot — A wood volume measurement containing 1,728 cubic inches, such as a piece of wood measuring 1 foot on a side. A cubic foot of wood contains approximately 6 to 10 usable board feet of wood.

Cull — A tree or log of merchantable size but no market value because of serious defects.

Cutting Cycle — The planned time interval between major harvesting operations in a stand. The term usually is applied to uneven-aged stands. For example, a cutting cycle of 10 years means that a harvest would be carried out in a stand every 10 years.

DBH — The diameter of a tree at breast height (4-1/2 feet above the ground).

DIB (d.i.b.) — Diameter inside bark. Diameter of a log measured inside the bark at the small end of a log.

Deciduous Tree — A tree that loses its leaves during the winter.

Defect — The portion of a tree or log that is unusable for the intended product. Defects include decay, crook, and excessive limbiness.

Depletion — The process of claiming an income tax deduction to recover the basis in certain capital assets, including timber, at the time of sale or other disposal.

Depletion Allowance — The proportion of the adjusted basis for a capital asset that may be deducted from timber sale or other proceeds when figuring income taxes payable.

Depreciation — The process of claiming an income tax deduction to recover the basis in certain capital assets that have a determinable, useful life, including equipment and buildings, over a fixed number of years.

Diameter — *See DBH or DIB.*

Discount Rate — Annual percentage rate by which future cash flows are reduced to make them equivalent to values in the base year.

Discounting — The process of determining the present value of a future payment.

Dominant — *See Crown Classification.*

Economies of Scale — The relationship between size and cost in which increasing the size of the producing entity and the total market share for the goods or increasing the number of units produced results in a reduction in per-unit production costs. Economies of scale also may result when the number of units purchased increases.

Ecosystem — An interacting system of plants, animals, microorganisms, soil, and climate.

Effective Interest Rate — The interest rate that is actually paid or earned. It considers multiple compounding periods within a year.

Environment — The prevailing conditions that reflect the combined influence of climate, soil, topography, and biological (other plants and animals) factors in an area.

Epicormic Branch — A branch that develops from an adventitious or dormant bud beneath the bark, usually on hardwoods. Epicormic branches often develop on tree trunks and major limbs after a stand has been heavily thinned. Also called water sprout.

Equivalent Annual Income (EAI) — The net present value converted to an annual value paid at the end of the year over the life of the investment with interest calculated at the appropriate discount rate. It is assumed that the investment would return the calculated amount annually. An EAI that is greater than or equal to zero indicates that the investment may be worth undertaking.

Even-Aged Forest — A forest or stand in which the age difference between trees forming the main canopy does not exceed 20 percent of the age of the stand at maturity.

Evergreen Tree — A tree that retains some or all of its leaves throughout the entire year, e.g., red pine, white spruce, white-cedar.

Fair Market Value — The price at which property would change hands between a buyer and seller, neither being required to buy or sell, and both having reasonable knowledge of all the necessary facts.

Financial Analysis — An analysis that estimates the profitability for an investment from the point of view of the decision maker(s) involved in the investment.

Foliage — Leaves on a tree or other plant.

Forest — A plant community in which the dominant vegetation is trees and other woody plants.

Forest Management — The process of giving a forest care so that it remains healthy and vigorous and provides the products and amenities the landowner desires; also, the application of technical forestry principles and practices and business techniques to the management of a forest.

Forest Type — A group of tree species which, because of their environmental requirements and tolerance for shade and moisture, often are found growing together. Examples are the jack pine type and the aspen-paper birch type.

Forestry — The science, art, and practice of managing trees and forests and their associated resources for human benefit.

Frilling — Completely encircling the trunk of a tree with axe cuts that sever the bark and cambium (actively growing layer of cells) with the intent of killing the tree. A herbicide often is injected into the frill.

Fungicide — A chemical that kills fungi.

General Inflation Rate — An interest rate used to incorporate changes in purchasing power within the economy.

Girdling — Completely encircling the trunk of a tree with a cut that severs the bark and cambium (actively growing layer of cells) and usually penetrates into the sapwood to kill the tree by preventing the conduction of water and nutrients.

Grading — Evaluating and sorting trees or logs according to quality.

Habitat — The local environment in which a plant or animal lives.

Hardboard — A panel of wood formed from wood fibers compressed together under heat.

Harden Off — The natural process a plant goes through late in the growing season to prepare it to survive winter weather. It involves thickening of cell walls and other physiological changes.

Hardwood — A term used to describe broadleaf, usually deciduous, trees such as oak, maple, ash, and elm. The term does not necessarily refer to the hardness of the wood.

Harvest — The felling and removal of final crop trees on an area to 1) obtain income, 2) develop the environment necessary to regenerate the forest, or 3) achieve objectives such as development of wildlife habitat. (Contrast with intermediate cut.)

Height, Merchantable — The height of a tree (or length of its trunk) up to which a particular product may be obtained. For example, if the minimum usable diameter of pulpwood sticks is 4 inches, the merchantable height of a straight tree would be its height up from stump height to a trunk diameter of 4 inches.

Height, Total — Height of a tree from ground level to the top of its crown.

Herb — A nonwoody plant.

Herbicide — A chemical that kills herbaceous (nonwoody) plants. In common usage, often used interchangeably with phytocide (plant killer) and silvicide (tree killer).

Increment Borer — A hollow, auger-like instrument used to bore into the trunk of a tree to remove a cylinder of wood containing a cross-section of the tree's growth rings.

Independent — The situation that exists where two or more alternatives can be chosen out of the candidate list; each alternative is evaluated on its own merit. Comparisons of independent investment proposals are designed to determine which alternatives satisfy some minimum set of qualifications; all of the alternatives that satisfy this minimum set could be implemented.

Insecticide — A chemical that kills insects.

Interest — The cost of using credit or another's money per period of time, expressed as a percentage.

Intermediate — *See Crown Classification And Shade Tolerance.*

Intermediate Cut — The removal of physically or financially immature trees from a stand after establishment and before a major harvest with the primary objective of improving the quality or growth of remaining trees. Contrast with Harvest. An intermediate cut may generate income (commercial cutting) or may cost the forest landowner (a noncommercial cutting).

Internal Rate of Return (IRR) — The rate at which the present value of an investment's present and future income just equals the present value of its present and future costs. This rate is found by a process of trial and error. When the IRR is greater than the MARR, the investment may be worth undertaking.

Investment Alternative — A scheme for investment resources that can be evaluated as an independent unit, separable from other investment options. Each alternative consists of costs, benefits, and a transformation function.

Investment Monitoring — A process of collecting further data and performing appropriate analyses after implementing an investment alternative in order to determine whether to continue, change, or liquidate the investment.

Kerf — The width of a saw blade. The cut made by a saw blade. The wood converted to sawdust as a saw cuts lumber from a log.

Layering — Process of regenerating a tree by covering a lower branch with soil or organic matter, after which the branch develops roots and can stand alone as a new tree.

Liberation — *See Release Cutting.*

Live-Crown Ratio — The percentage of total tree height that has live branches on it (see Figure 22).

Log — A cut piece of the woody stem of a tree. The trunk portion of a tree. A 16-foot long piece of a tree stem.

Log Rule — A device, usually a table, that estimates log volume based on log diameter (inside bark at the small end) and length. A log rule expresses the volume of cut logs. Contrast with tree rule.

Logger — An individual whose occupation is cutting timber.

Lump-Sum Sale — A technique of selling timber in which a single sum of money is paid for all of the timber regardless of the amount actually harvested. It is distinguished from a scale or unit sale, in which payment is based on the amount harvested (e.g., so much per cord or thousand board feet).

Management — *See Forest Management.*

Marginal Tax Rate — An income tax rate applied to income within a specified range.

Mast — Tree-produced nuts (hard mast) and fleshy fruits (soft mast) that are edible to wildlife.

Mature Tree — A tree that has reached the desired size or age for its intended use.

Measure of Investment Worth — Investment criterion used to allow comparison of the financial profitability of different investment alternatives.

Minimum Acceptable Rate of Return (MARR) — The return (an interest rate) an investor requires to commit any monies to an investment at a given level of risk. Given equal risk, an investment alternative is unacceptable unless its expected return equals or exceeds the MARR.

Multiple Use Management — The management of land for more than one purpose.

Mutually Exclusive — The situation where only one of the feasible candidate alternatives can be chosen; that is, if one alternative is chosen, all others are excluded.

Natural Forest Stand — A forest stand that originates from seed, seedlings, root suckers, or stump sprouts that are naturally present on the site.

Net Present Value (NPV) — The present value of an investment's benefits minus the present value of its costs, all discounted at the appropriate discount rate. An NPV greater than or equal to 0 indicates that the investment may be worth undertaking.

Noncommercial Cutting — A cutting that does not yield a net income, usually because the trees harvested are too small, of poor quality, or of nonmerchantable species.

Overstory — The highest canopy in a stand of trees. Contrast with Understory.

Period — The length of one financial analysis interval, frequently 1 year.

Photosynthesis — A chemical reaction that takes place in green plants in which carbon dioxide and water combine to produce sugar.

Phytocide — A chemical that kills plants.

Plantation — An artificially reforested area established by planting or direct seeding.

Planting Stock — Seedlings or transplants that are planted to reproduce a tree stand.

Plot — An area of land, usually less than one acre, on which trees and sometimes other vegetation are measured during a cruise (or inventory).

Pole Stand — A stand of trees where DBH ranges from 4 inches to approximately 8 to 12 inches.

Poletimber — *See Pole Stand.*

Precommercial Cutting — *See Noncommercial Cutting.*

Principal — The money to which compounding is applied or against which interest is charged.

Pruning — The removal of live or dead branches from standing trees. With forest trees, pruning generally means removing limbs from the lower 17 feet of the main stem to produce higher quality (knot-free) wood.

Pulpwood — Wood cut primarily to be converted into wood pulp, chips, or fiber for the manufacture of paper, fiberboard, or other wood fiber products.

Range — The geographic area in which a tree species grows. Natural range is the geographic area where a species is known to occur under natural conditions; commercial range is the geographic area in which a species is harvested for commercial purposes.

Real Price Change — A change in the real purchasing power or value of an activity beyond the general inflation rate.

Reforestation — Reestablishing a forest on an area where forest vegetation has been removed.

Regeneration — *See Reproduction.*

Release Cutting — A cutting operation carried out to release young trees (seedlings or saplings) from competition with other trees of the same size (a "cleaning") or from larger and overtopping trees (a "liberation").

Reproduction — The process by which the forest is replaced or renewed by natural or artificial means. Also, the young crop itself.

Risk — The chance of a loss. In some cases, the risk can be estimated using probability principles. In other cases, the risk cannot be quantified. The terms risk and uncertainty are sometimes used interchangeably.

Root Collar — The place on a tree stem at or slightly below the ground line where roots first appear and the stem usually is slightly swollen.

Root Sucker — Above-ground shoot growth that originates from buds on the lateral roots of a tree. Root suckers usually develop when the parent tree dies or is severely damaged.

Roots — That portion of the tree that generally is underground and that functions in nutrient absorption, anchorage, and storage of food and waste products.

Rotation — The number of years required to establish and grow trees to a specified size, product, or maturity.

Salvage Cut — Harvesting trees that have been killed or are in danger of being killed by insects, disease, or the environment to save their economic value.

Sanitation Cut — The harvesting or destruction of trees infected or highly susceptible to insects or diseases to protect the rest of the stand.

Sapling — A small tree; often defined as being between 1 and 4 inches DBH.

Sawlog — A log large enough to produce a sawn product—usually at least 10 to 12 inches in diameter, 8 feet long, and solid.

Sawtimber — Standing trees large enough to produce sawlogs.

Scale Stick — A flat stick, similar to a yardstick, that is calibrated so log volumes can be read directly when the stick is placed on the small end of a log of known length.

Scaling — Process of measuring wood products, usually pulpwood and sawlogs, after the trees are felled.

Scalping — Removing a patch or strip of sod to expose mineral soil in preparation for planting trees.

Scarification — Churning the soil surface to expose mineral soil and uproot vegetation to prepare a seedbed for natural or artificial seeding.

Seed Cut — A harvest in a shelterwood system that is designed to encourage the growth of desirable seed-producing trees and control undesirable understory trees, shrubs, and herbaceous plants with the overall goal of encouraging seed production, seed germination, and seedling survival.

Seed Tree — A tree left standing after a timber harvest as a source of seed for reproducing a new timber stand.

Seed Tree Harvest — A harvest in which all of the trees are removed from the harvest area except for a few scattered trees that are left to provide seed to establish a new forest stand. Produces an even-aged stand.

Seeding — Artificially scattering tree seeds over an area to establish a stand.

Seedling — A tree, usually defined as less than 1 inch in DBH, that has grown from a seed (in contrast to a sprout).

Selection Harvest — A harvest in which individual trees or small groups of trees are harvested at periodic intervals (usually 8 to 15 years) based on their physical condition or degree of maturity. Produces an uneven-aged forest.

Sensitivity Analysis — An assessment of the impact that changes in assumptions will have on the profitability of an investment. This analysis is conducted because there is uncertainty in cash flow and interest rate assumptions, particularly in the distant future.

Shade Tolerance — Relative ability of a tree species to reproduce and grow under shade. Tree species usually are classified in descending order of shade tolerance as very tolerant, tolerant, intermediate, intolerant, or very intolerant.

Shelterwood Harvest — A harvest in which trees are removed in a series of two or more cuttings to allow the establishment and early growth of new seedlings under the partial shade and protection of older trees. Produces an even-aged forest.

Shrub — A low-growing perennial plant with a persistent woody stem or stems and low-branching habit.

Silvicide — A chemical that kills woody plants.

Silviculture — The art, science, and practice of establishing, tending, and reproducing forest stands of desired characteristics based on knowledge of species characteristics and their environmental requirements.

Site — A contiguous area selected for its capacity to produce a particular forest or other vegetation based on the combination of biological, climatic, and soil factors present.

Site Index — An expression of forest site quality based on the expected height of dominant and codominant trees at a specified age (in the Midwest, usually 50 years).

Site Preparation — Preparing an area of land by clearing, chemical vegetation control, burning, etc., for the purpose of establishing a new stand of trees.

Slash — Residue, including tree tops, branches, bark, and unmerchantable wood, left on the ground after logging, pruning, or other forest operations.

Snag — A standing dead tree.

Softwoods — *See Conifer.*

Soil Expectation Value (SEV) — The value of bare land, equivalent to the present value of an infinitely long series of cash flows resulting from land management. It represents the maximum amount an investor could pay for a tract of land and still earn the required interest rate. An SEV greater than or equal to zero or the selling price of the land indicates that the investment may be worth undertaking.

Soil Texture — The feel or composition of a soil based on the proportion of sand, silt, and clay.

Species — A recognizable collection of plants or animals (e.g., sugar maple or red pine) that are so similar they suggest a common parentage and produce like offspring.

Species Composition — The mix of tree species occurring together in a stand.

Sprout — A tree that has grown from the stump of another tree.

Stand — A group of trees occupying a given area and sufficiently uniform in species composition, tree size distribution, stocking, and soil characteristics so as to be distinguishable from the adjoining forest. A forest stand is said to be **pure** if 80 percent or more of the trees present are of the same species. Otherwise, the stand is said to be **mixed**.

Stand Density — *See Stocking.*

Stocking — A measure of the degree of crowding of trees in a stand, also known as stand density. Commonly expressed by the number of trees per acre or percent of crown cover.

Structural Board — A wood panel made from chips or flakes that have been formed into a panel by heat, pressure, and sometimes an adhesive. Frequently used in construction for underlayment on floors, roofs, and walls.

Stumpage — The dollar value of a standing tree or group of trees.

Succession — The process by which one plant community is gradually replaced by another.

Sucker — *See Root Sucker.*

Suppressed — *See Crown Classification.*

Sweep — The curvature in a log that gives the log a slight C-shape.

TSI (Timber Stand Improvement) — The practice of removing undesirable trees, shrubs, vines, or other vegetation to achieve the desired stocking of the best quality trees.

Thinning — Cutting trees in an immature forest stand to reduce the stocking and concentrate growth on fewer, higher quality trees.

Timber — Standing trees, usually of commercial size.

Timber Inventory — A collection of information about a timber stand made by measuring tree and stand characteristics such as tree volume and grade and stand density.

Time Value of Money — Concept that reflects the fact that cash flows received or paid in future years do not have the same value as those received or paid today. Future cash flows are larger due to allowances for factors such as uncertainty, risk, and inflation. It is the concept underlying compounding.

Tolerance — *See Shade Tolerance.*

Tract — A contiguous area of land (and water).

Transplant — A tree seedling that was transplanted at least once in the nursery.

Tree — A woody plant having a well-defined stem and a more or less definitely formed crown and which usually attains a height of at least 10 feet.

Tree Farm — A privately owned woodland dedicated to the production of timber crops. It may be formally recognized by the Tree Farm program of the American Forest Council.

Tree Injector — Equipment specially designed to inject chemicals into the trunk of a tree to kill it.

Tree Rule — A device, usually a table, that estimates the volume of standing trees based on diameter (DBH) and height.

Trunk — Main stem or bole of a tree.

Understory — Low-growing woody or herbaceous vegetation that forms a layer beneath the overstory.

Uneven-Aged Forest — A forest or stand in which there are more than two age classes of trees present. There usually is a minimum age difference of 20 percent of the rotation length in years.

Veneer — Thin sheets of wood (usually less than 1/4" thick) produced by slicing or peeling a log. Also used in reference to a tree or log suitable for cutting into veneer because of its species, size, and good quality.

Volume Table — A table that estimates the volume of wood in a standing tree based on measurements of the tree (most commonly DBH and merchantable height).

Water Table — The highest point in a soil profile at which water saturates the soil on a seasonal or permanent basis.

Wolf Tree — A tree that occupies more space in the forest than its timber value justifies. Usually a tree that is older and has a larger crown than other trees in the stand.

Wood Pulp — Mechanically ground or chemically digested wood (composed primarily of wood fiber) used in the manufacture of paper.

Woodland Management — *See Forest Management.*

SUBJECT INDEX

Note: page references to illustrations are in italics

A

B

C

D

E

F

G

H

I

J

K

L

M

Q

R

S

T

U

V

W

Y

Z